NUMERICAL METHODS OF
MATHEMATICAL OPTIMIZATION

Computer Science and Applied Mathematics

A SERIES OF MONOGRAPHS AND TEXTBOOKS

Editor
Werner Rheinboldt
University of Maryland

Hans P. Künzi, H. G. Tzschach, and C. A. Zehnder
NUMERICAL METHODS OF MATHEMATICAL OPTIMIZATION: WITH ALGOL
AND FORTRAN PROGRAMS, CORRECTED AND AUGMENTED EDITION, 1971

Azriel Rosenfeld
PICTURE PROCESSING BY COMPUTER, 1969

James Ortega and Werner Rheinboldt
ITERATIVE SOLUTION OF NONLINEAR EQUATIONS IN
SEVERAL VARIABLES, 1970

A. T. Berztiss
DATA STRUCTURES: THEORY AND PRACTICE, 1971

Azaria Paz
INTRODUCTION TO PROBABILISTIC AUTOMATA, 1971

In preparation

David Young
ITERATIVE SOLUTION OF LARGE LINEAR SYSTEMS

Ann Yasuhara
RECURSIVE FUNCTION THEORY AND LOGIC

Numerical Methods of Mathematical Optimization

With ALGOL and FORTRAN Programs

CORRECTED AND AUGMENTED EDITION

HANS P. KÜNZI
*University of Zürich
and Eidgenössische Technische Hochschule
Zürich, Switzerland*

H. G. TZSCHACH
*Division of Mathematical Methods
IBM Deutschland, Berlin, Germany*

C. A. ZEHNDER
*Eidgenössische Technische Hochschule
Zürich, Switzerland*

Translated by Werner C. Rheinboldt
*Institute for Fluid Dynamics and Applied Mathematics
University of Maryland, College Park, Maryland*

and Cornelie J. Rheinboldt

ACADEMIC PRESS
NEW YORK and LONDON   1971

ACADEMIC PRESS, INC.
111 Fifth Avenue, New York, New York 10003

United Kingdom Edition published by
ACADEMIC PRESS, INC. (LONDON) LTD.
Berkeley Square House, London W1X 6BA

LIBRARY OF CONGRESS CATALOG CARD NUMBER: 68-18673

PRINTED IN THE UNITED STATES OF AMERICA

Numerical Methods of Mathematical Optimization
With ALGOL and FORTRAN Programs

First published in the German language under the title *Numerische Methoden der mathematischen Optimierung* (*mit ALGOL und FORTRAN Programmen*) and copyrighted in 1966 by B. G. Teubner Verlag, Stuttgart, Germany.

This is the only authorized English edition published with the consent of the publishing house B. G. Teubner, Stuttgart, of the original German edition which has been published in the series "Leitfäden der angewandten Mathematik und Mechanik," edited by Professor Dr. H. Görtler.

PRINTED IN THE UNITED STATES OF AMERICA

PREFACE TO THE GERMAN EDITION

Compared to the already existing literature on linear and nonlinear optimization theory, this book differs both in content and presentation as follows:

The first part (Chapters 1 and 2) is devoted to the theory of linear and nonlinear optimization, with the main stress on an easily understood, mathematically precise presentation. The chapter on linear optimization theory is somewhat more detailed than the one on nonlinear optimization. Besides the theoretical considerations, several algorithms of importance to the numerical application of optimization theory are discussed. As prerequisite mathematical knowledge only the fundamentals of linear algebra, predominantly vector and matrix algebra, and the elements of differential calculus are required of the reader (the latter for the nonlinear optimization).

One difference between our presentation and earlier ones will undoubtedly be the fact that in the second part we have developed both an ALGOL and a FORTRAN program for each one of the algorithms described in the theoretical section. This should result in easy access to the application of the different optimization methods for everyone familiar with these two symbolic languages. (The difference between the ALGOL and FORTRAN programs is in the language only—computationally they proceed entirely in parallel.)

Intentionally, both parts have been kept largely independent of each other so that the first part can be used as an independent theoretical presentation and the second part as a well-rounded and efficient program collection. The connection between the theory and the programs is assured by an intermediary text (Chapter 3) which also contains all those explanations needed for the use of ALGOL and FORTRAN programs.

v

The first author listed carries the principal responsibility for the theoretical part, the other two prepared the program section.

We hope that with this division into two separate parts we have succeeded in bringing the theory and the practical application of the optimization methods closer to each other. Anyone working in this subject area knows that without electronic computers, and therefore without computer programs, the actual application of linear and nonlinear optimization belongs more or less to the realm of a utopia.

We wish to thank Professors E. Stiefel and H. Görtler as well as Drs. P. Kall and Kirchgässner for their many suggestions and valuable recommendations for improvements. Mr. D. Latterman has assisted us greatly with the programming, and valuable assistance was also given us in our proofreading by Drs. Kleibohm and Tan. We should furthermore like to express our warm thanks to the publishers for their consideration of our numerous wishes and for their careful printing job.

H. P. KÜNZI
H. G. TZSCHACH
C. A. ZEHNDER

Autumn 1966
Zürich and Berlin

CONTENTS

3. Explanations of the Computer Programs

4. ALGOL and FORTRAN Programs

1 LINEAR OPTIMIZATION

1.1 General Formulation of Linear Optimization

Linear optimization concerns the optimization of a linear expression subject to a number of linear constraints, and can involve either a maximization or a minimization problem.

First, we will formulate the maximization problem. In that case, quantities $x_1, x_2, \ldots, x_n$ are to be found for which the linear form, or objective function,

$$B = \sum_{i=1}^{n} a_{0i} x_i$$

assumes a maximum subject to the constraints

$$\sum_{i=1}^{n} a_{ji} x_i \leq a_{j0} \qquad (j = 1, 2, \ldots, m) \tag{1.1}$$

as well as the nonnegativity restrictions

$$x_i \geq 0 \qquad (i = 1, 2, \ldots, n).$$

Here the coefficients a_{0i}, a_{ji}, and a_{j0} are given and the x_i are the original unknown variables.

It will be useful for further discussion to transform the system of constraints in (1.1) into a system of equations of the form

$$\sum_{i=1}^{n} a_{ji} x_i + x_{n+j} = a_{j0} \qquad (j = 1, 2, \ldots, m)$$

by introducing additional variables (so-called slack-variables) which have to satisfy the additional nonnegativity restrictions

$$x_{n+j} \geq 0 \qquad (j = 1, 2, \ldots, m).$$

Using matrix and vector notation, we can then represent the maximization problem in the following shortened form:

Maximize the objective function

$$B = a_{0.}^T x$$

subject to the constraints

$$Ax = a_{.0}$$

$$x \geq 0$$

$$\left.\begin{array}{c} \\ \\ \\ \end{array}\right\} \quad (1.2)$$

In contrast to (1.1) the symbols used in form (1.2) have been slightly modified.

x^T is a row vector with $(m + n)$ components

$$x^T = (x_1, x_2, \ldots, x_n, \quad x_{n+1}, \ldots, x_{n+m})$$

and the row vector $a_{0.}^T$ also has $(m + n)$ components

$$a_{0.}^T = (a_{01}, a_{02}, \ldots, a_{0n}, \quad a_{0,n+1}, \ldots, a_{0,n+m})$$

where

$$a_{0,n+1} = a_{0,n+2} = \cdots = a_{0,n+m} = 0.$$

A is an $m \times (m + n)$ matrix, given by

$$A = \begin{pmatrix} a_{11} & a_{12} & \cdots & a_{1n} & 1 & 0 & \cdots & 0 \\ a_{21} & a_{22} & \cdots & a_{2n} & 0 & 1 & \cdots & 0 \\ \cdot & & & & & & & \\ \cdot & & & & & & & \\ \cdot & & & & & & & \\ a_{m1} & a_{m2} & \cdots & a_{mn} & 0 & 0 & \cdots & 1 \end{pmatrix}. \quad (1.3)$$

The constraint system in (1.2) consists of m equations in the $(m + n)$ unknowns $x_1, x_2, \ldots, x_N$; $N = m + n$.

If A has rank m,[1] a set of n variables $x_{v_1}, \ldots, x_{v_n}$ can always be chosen from among the $m + n$ variables $x_1, \ldots, x_{m+n}$ such that when these $x_{v_1}, \ldots, x_{v_n}$ are given arbitrary but fixed values, the system

$$Ax = a_{.0} \quad (1.4)$$

is uniquely solvable in terms of the remaining m variables.

[1] The structure of the matrix A could also be more general than that shown in (1.3).

Any vector x for which the components satisfy both the system of constraints and the nonnegativity restrictions of (1.2) is called a feasible vector.

A feasible vector x^0 which at the same time maximizes the objective function is called an optimal feasible vector. Using these definitions we can formulate (without proof) the fundamental theorem of linear programming:

If an optimal feasible vector exists at all, there always exists an optimal feasible vector with at least n zero components.

If the row vectors of the matrix A, corresponding to the nonzero components of x, are linearly independent, x is called a basic vector.[1] In its strict form we then have the

Fundamental Theorem of Linear Optimization. *If an optimal feasible vector exists, there also exists a feasible basic vector which is optimal.*

It is easily seen that the first formulation is derived from the second one. A proof of this fundamental theorem can be found in 1.4. In the event only two basic variables ($n = 2$) occur in system (1.2), the result can be described graphically.

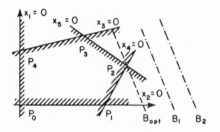

Fig. 1. Geometric interpretation of a linear optimization problem in the plane.

Suppose the constraint system in (1.2) has three equations, i.e., that $m = 3$ and $n = 2$, then a geometrical interpretation follows from Fig. 1.

The first inequality, to which the slack variable x_3 belongs, is satisfied by all points contained in the hatched half-plane bordered by the straight

[1] *Translator's Comment:* The term "basic solution" is also used and the associated nonzero variables are called the basic variables. In view of the fact that in the nondegenerate case they correspond to a basis of the space, the set of basic variables is also called the basis for short.

line $x_3 = 0$. The same holds for x_4 and x_5. Two other half-planes are determined by the nonnegativity restrictions $x_1 \geq 0$, $x_2 \geq 0$, and the intersection of all these half-planes constitutes the convex domain $P_0 P_1 P_2 P_3 P_4$.

If we now consider the objective function for different values B_i, we are obviously faced with the problem of finding the "outermost corner" of the convex polygon $P_0 P_1 P_2 P_3 P_4$ with respect to the family of straight lines B_i.

This corner is represented by a basic solution[1] because in the case of Fig. 1 the optimal feasible solution vector

$$x^T = (x_1, x_2, x_3, x_4, x_5)$$

for the cornerpoint P_2 has the positive components x_1, x_2, and x_3 while x_4 and x_5 are zero.

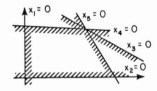

Fig. 2. Geometrical interpretation of a degenerate linear optimization problem in the plane.

The different methods of linear optimization are primarily directed toward finding an efficient algorithm for calculating as quickly as possible the "outermost corner" P_2 drawn in Fig. 1. One of the main purposes of this book has been to describe such algorithms from both a theoretical as well as a numerical standpoint. If the point of solution (for $n = 2$) lies in the intersection of 3 or more straight lines (Fig. 2), we speak of a degenerate linear optimization problem. In anticipation of our further discussion, it should be noted that in the general case when $n \geq 2$, a degenerate solution has more than n vanishing variables; for $n = 2$ this is evident from Fig. 2 (see also 1.5).

As was mentioned in the beginning, we can also consider minimization problems. Every maximization problem can be trivially changed into a minimization problem, and for problem (1.1) this means the following:

[1] See Karlin [82, Chapter 6.1, pp. 161–162].

Minimize

$$-B = -\sum_{i=1}^{n} a_{0i} x_i$$

subject to the constraints

$$-\sum_{i=1}^{n} a_{ji} x_i \geq -a_{j0} \qquad (j = 1, \ldots, m)$$

and the nonnegativity restrictions

$$x_i \geq 0 \qquad (i = 1, \ldots, n).$$

(1.5)

Algorithms must also be found for the minimization problems permitting the determination of an optimal feasible solution. It is up to the reader to depict the minimization problem graphically in the case $n = 2$.

For reasons which will become clear in connection with the duality principle of linear optimization (see 1.6), it is desirable to construct the minimization problem in a way which is symmetrical to the maximization problem (1.1).

Minimize

$$C = \sum_{j=1}^{m} a_{j0} w_j$$

subject to the constraints

$$\sum_{j=1}^{m} a_{ji} w_j \geq a_{0i} \qquad (i = 1, 2, \ldots, n)$$

and the nonnegativity restrictions

$$w_j \geq 0 \qquad (j = 1, 2, \ldots, m).$$

(1.6)

Using vector and matrix notation as was done in problem (1.2), this can be stated as follows:

Minimize

$$C = a_{.0}^T w$$

subject to the constraints

$$A^T w = a_0.$$

$$w \geq 0$$

(1.7)

where
$$w^T = (w_1, w_2, \ldots, w_{m+n})$$

$$A^T = \begin{pmatrix} a_{11} & a_{21} & \cdots & a_{m1} & -1 & 0 & \cdots & 0 \\ a_{12} & a_{22} & \cdots & a_{m2} & 0 & -1 & \cdots & 0 \\ \cdot & & & & & & & \\ \cdot & & & & & & & \\ \cdot & & & & & & & \\ a_{1n} & a_{2n} & \cdots & a_{mn} & 0 & 0 & \cdots & -1 \end{pmatrix}$$

In (1.6) it will be proved that for the solution of the problems (1.1) and (1.6) the following fundamental rule holds:

$$B_{\max} = C_{\min}. \tag{1.8}$$

1.2   The Simplex Method

Starting from the problem formulation (1.2) we shall now try to find an algorithm which yields the desired optimal value of the objective function. It will be assumed that the linear expressions on the left side of the constraints in (1.2) are linearly independent. For the time being we shall furthermore suppose that the coefficients a_{j0} are nonnegative. This restriction will be dropped later.

One of the best known methods for calculating the optimal solution of (1.2) is the so-called simplex method, first published by Dantzig [22] in 1948.

This iterative method is derived from the fundamental theorem (cf. 1.1). The first step is to look for a nondegenerate basic solution[1] of the system (1.2), that is, a basic solution with exactly n zero variables.

If this basic solution has at least one component $x_i < 0$, it is not feasible and has to be dropped from further consideration. However, if the solution is feasible (i.e., if $x_i \geq 0$ for all i), it serves as the starting point for the following iterative process, provided, of course, that this process is still required. As we shall prove in 1.4, the simplex method leads in finitely many steps to the optimal solution.

Suppose we are at the kth step of the iteration but that the optimum has not been reached as yet. We then proceed to the $(k + 1)$ step by letting one variable become positive which at the kth step was zero, i.e.,

---

[1] The case of degenerate basic solutions will be discussed in 1.5.

which did not belong to the basis, and, in turn, by setting another variable equal to 0 which belonged to the basis at the kth step. In other words, each iteration step consists of a so-called exchange of variables where one variable x_e is entering the basis and another variable x_f is leaving it.

The simplex method provides the rules for selecting the variables for the exchange in such a way that the objective function undergoes an improvement in the direction of the optimum. The method also provides us with a criterion for recognizing when the optimum has been reached.

If in (1.1) or (1.2) all a_{j0} are positive, it is often useful to set all original variables $x_1, x_2, \ldots, x_n$ equal to 0 for the first step of the iteration. This means that the slack variables $x_{n+1}, x_{n+2}, \ldots, x_{n+m}$ form the first basis.

The system of equations in (1.2) is now solved with respect to these basic variables:

$$x_{n+j} = a_{j0} + \sum_{i=1}^{n} a_{ji}(-x_i) \geq 0 \qquad (j = 1, \ldots, m). \tag{1.9}$$

If the value of the objective function (1.2) for this basic solution is denoted by a_{00}, then

$$a_{00} = \sum_{j=1}^{m} a_{0,n+j} x_{n+j} = \sum_{j=1}^{m} a_{0,n+j} a_{j0}$$

Considering (1.9), we then have for the objective function

$$z = \sum_{i=1}^{n} a_{0i} x_i + \sum_{j=1}^{m} a_{0,n+j} x_{n+j} = \sum_{i=1}^{n} a_{0i} x_i + \underbrace{\sum_{j=1}^{m} a_{0,n+j} a_{j0}}_{=a_{00}} + \sum_{j=1}^{m} \sum_{i=1}^{n} a_{0,n+j} a_{ji}(-x_i)$$

and therefore

$$z = a_{00} + \sum_{i=1}^{n} \left(\sum_{j=1}^{m} a_{ji} a_{0,n+j} - a_{0i} \right)(-x_i) = a_{00} + \sum_{i=1}^{n} \alpha_{0i}(-x_i) \tag{1.10}$$

where

$$\alpha_{0i} = \sum_{j=1}^{m} a_{ji} a_{0,n+j} - a_{0i}. \tag{1.11}$$

We shall now explain the iterative process step by step, although several of the proofs will only be detailed further in 1.4.

STEP 1. The system of Eqs. (1.9) and the objective function (1.10) have been written in the form of a table (1.12), called a tableau.

The quantities α_{0i} in the bottom-row are called valuation coefficients. They correspond to the coefficients of the objective function (1.10).

		$-x_1$	$-x_2 \cdots -x_n$	$-x_{n+1}$	$-x_{n+2} \cdots -x_{n+m}$
x_{n+1}	a_{10}	a_{11}	$a_{12} \cdots a_{1n}$	1	0 $\cdots$ 0
x_{n+2}	a_{20}	a_{21}	$a_{22} \cdots a_{2n}$	0	1 $\cdots$ 0
.	.	.			
.	.	.			
x_{n+m}	a_{m0}	a_{m1}	$a_{m2} \cdots a_{mn}$	0	0 $\cdots$ 1
z	a_{00}	α_{01}	$\alpha_{02} \cdots \alpha_{0n}$	0	0 $\cdots$ 0

$$(1.12)$$

STEP 2 (Check for Optimality). In order to check whether the solution given in the second column of tableau (1.12) is already optimal, we calculate by (1.11) the valuation coefficients α_{0i} for each variable x_i not in the basis:

$$\alpha_{0i} = \sum_{j=1}^{m} a_{ji} a_{0,n+j} - a_{0i} \qquad (i = 1, \ldots, n).$$

The absolute values of the negative valuation coefficients α_{0i} indicate by how many units the objective function would be increased if the feasible solution just obtained were subjected to the following change: The (zero) variable x_i is increased from 0 to 1, the other (zero) variables are retained at their zero value, and then the basic variables (nonzero variables) are modified such that the constraints remain satisfied.

If one or more of the valuation coefficients α_{0i} are negative, the solution cannot be optimal yet and at least one additional iteration step is required.

If all α_{0i} are positive, the optimal solution has been reached. If, besides strictly positive α_{0i}, some of these quantities have the value zero, a number of equivalent optimal solutions exist (in this connection see the proofs in 1.4).

STEP 3 (Determination of the Variable x_e Entering the Basis). Search for a variable with an index i with the properties
(1) $x_i = 0$ in the tableau (1.12)
(2) $\alpha_{0i} < 0$
(3) at least one of the a_{ij} ($j = 1, \ldots, m$) in the x_i-column is greater than zero.

On the basis of the definition of the α_{0i} in (1.11) it is reasonable to choose that variable x_i as the one to enter the basis for which α_{0i} is most negative; in other words, for which $\max_i (-\alpha_{0i})$ is assumed.

If condition (3) is not satisfied, a_{00} will grow indefinitely when the corresponding x_i is increased arbitrarily (in this connection see also Charnes et al. [20]).

STEP 4 (Determination of the Variable x_f Leaving the Basis). x_f is determined in such a way that none of the new basic variables will become negative. The corresponding criterion follows immediately from an explicit computation. Determine the index f in such a way that the quotient a_{j0}/a_{je}, formed for all j with $a_{je} > 0$, assumes its smallest value.

In the matrix A of the tableau (1.12), the column belonging to x_e is called the pivot column and, correspondingly, the row belonging to x_f the pivot row. For the subsequent transformation of the tableau the pivot element a_{fe} is of particular significance.

STEP 5 (Transformation of Tableau (1.12)). The transformation rules leading from one tableau to the next represent elementary row operations for the matrix of (1.12). These row operations are selected in such a way that the transformed pivot element a'_{fe} will equal 1 while all other elements in the pivot column assume the value 0:

		$-x_1$	$-x_2 \cdots -x_e \cdots -x_n$			$-x_{n+1}$	$-x_{n+2}- \cdots -x_f \cdots -x_{n+m}$		
x_{n+1}	a'_{10}	a'_{11}	$a'_{12} \cdots$	0	$\cdots \ a'_{1n}$	1	0	$\cdots \ a'_{1f} \cdots$	0
x_{n+2}	a'_{20}	a'_{21}	$a'_{22} \cdots$	0	$\cdots \ a'_{2n}$	0	1	$\cdots \ a'_{2f} \cdots$	0
.									
.									
.									
x_f	a'_{f0}	a'_{f1}	$a'_{f2} \cdots$	1	$\cdots \ a'_{fn}$	0	0	$\cdots \ a'_{ff} \cdots$	0
.									
.									
.									
x_{n+m}	a'_{m0}	a'_{m1}	$a'_{m2} \cdots$	0	$\cdots \ a'_{mn}$	0	0	$\cdots \ a'_{mf} \cdots$	1
z	a'_{00}	α'_{01}	$\alpha'_{02} \cdots$	0	$\cdots \ \alpha'_{0n}$	0	0	$\cdots \ \alpha'_{0f} \cdots$	0

$$(1.13)$$

Computationally this means the following:

i. Divide all elements of the pivot row by the pivot element a_{fe}.

ii. Then, in order to obtain the zeros in the other places of the pivot column, subtract suitable multiples of the (new) pivot row from the remaining $(m - 1)$ rows of tableau (1.12).

The selection of x_e and x_f assures that all a'_{j0} will turn out to be non-negative. The elements a'_{ji} and a'_{j0} not belonging to the pivot row are calculated in line with rule ii by

$$a'_{ji} = a_{ji} - a_{fi}\frac{a_{je}}{a_{fe}}$$

$$a'_{j0} = a_{j0} - a_{f0}\frac{a_{je}}{a_{fe}}.$$

$$(1.14)$$

Following this, the columns of x_e and x_f in (1.13) are interchanged and in the lead column x_f is replaced by x_e.

STEP 6.   Steps 2 through 5 are repeated until step 2 signifies that the optimum has been achieved.

Similar to (1.11) the current valuation quantities are calculated by:

$$\alpha'_{0[i]} = \sum_{[j]} \alpha'_{j[i]} a'_{0j} - a'_{0[i]} \tag{1.15}$$

where $[i]$ and $[j]$ extend over the indices of the nonbasic variables and the basic variables, respectively.

In many cases it is possible to determine a nontrivial basic solution for (1.2) ahead of time. Suppose it consists of the variables $x_1, x_2, \ldots, x_m$ with the values $a_{10}, a_{20}, \ldots, a_{m0}$, which means that $x_{m+1} = x_{m+2} = \cdots = x_{m+n} = 0$. In this case we begin with the system

$$\left. \begin{aligned} x_j &= a_{j0} + \sum_{i=1}^{n} a_{j,m+i}(-x_{m+i}) \geq 0 \qquad (j = 1, \ldots, m) \\ z &= a_{00} + \sum_{i=1}^{n} a_{0,m+i}(-x_{m+i}) \\ x_1 &\geq 0, \ldots, x_{m+n} \geq 0. \end{aligned} \right\} \tag{1.16}$$

As in the case of (1.12), this system (1.16) can again be written in the form of a table, and the rules for optimization remain the same as those given above. Frequently, while filling out a tableau, the unit vectors belonging to the basic variables are suppressed; this changes nothing in the computational rules, of course. For the last system (1.16) the reduced table (1.17) has the following form:

		$-x_{m+1}$	$-x_{m+2} \cdots - x_{m+n}$		
x_1	a_{10}	$a_{1,m+1}$	$a_{1,m+2}$	$\cdots$	$a_{1,m+n}$
x_2	a_{20}	$a_{2,m+1}$	$a_{2,m+2}$	$\cdots$	$a_{2,m+n}$
$\cdot$	$\cdot$	$\cdot$			
$\cdot$	$\cdot$	$\cdot$			
$\cdot$	$\cdot$	$\cdot$			
x_m	a_{m0}	$a_{m,m+1}$	$a_{m,m+2}$	$\cdots$	$a_{m,m+n}$
z	a_{00}	$\alpha_{0,m+1}$	$\alpha_{0,m+2}$	$\cdots$	$\alpha_{0,m+n}$

$$\tag{1.17}$$

For the sake of simplicity, the coefficients in system (1.16) have again been denoted by a_{ji} and a_{j0}, as in system (1.12), though these quantities do, of course, differ from each other since in general we obtain (1.16) from (1.2) only after several iteration steps.

Solution of the Minimization Problem. The solution of the minimization problem by means of the simplex method proceeds analogously to that of the maximization problem. Starting from the problem-formulation (1.7) we first search for a feasible solution as had been proposed for the minimization problem, after which a corresponding tableau is formed. Here, the trivial initial solution

$$x_1 = x_2 = \cdots = x_n = 0$$

and

$$x_{n+1}, x_{n+2}, \ldots, x_{n+m} \geq 0$$

will be feasible only in rare cases because, as easily seen, all coefficients of the vector a_0 in (1.7) would have to be nonpositive.

Once a feasible initial solution has been found, a tableau of the form (1.17) can be constructed and the iterative process applied with the same rules as those given above in steps 1–6; the only exception here is that in step 2 with its criterion for optimality the valuation coefficients have to show the opposite sign of those in the maximation problem:

STEP 2. In order to check whether the solution given in (1.17) is optimal, compute (by (1.11)) for each variable x_i not contained in the basis the valuation coefficients α_{0i}. If one or more of these quantities α_{0i} is positive, the solution cannot be optimal and at least one further iteration step is required. If all α_{0i} are nonpositive, the optimal solution has been reached.

Example 1. With the help of the simplex method maximize the objective function

$$B = 3x_1 + 5x_2 + 4x_3$$

subject to the constraints

$$2x_1 + 3x_2 \qquad \leq 8$$

$$2x_2 + 5x_3 \leq 10$$

$$3x_1 + 2x_2 + 4x_3 \leq 16$$

$$x_1 \geq 0, \qquad x_2 \geq 0, \qquad x_3 \geq 0.$$

Following the introduction of the nonnegative slack variables x_4, x_5, and x_6 form the initial tableau (1.17). The following iteration steps were calculated with the help of the computer programs given in Chapter 4, and in each case the second decimal was rounded off. The pivot element in each tableau is framed and the variables about to enter or exit are designated by arrows. A corresponding representation was also chosen for the examples that follow.

		$\downarrow$			
		$-x_1$	$-x_2$	$-x_3$	
$\leftarrow$ x_4	8	2	$\boxed{3}$	0	
x_5	10	0	2	5	1st basis: x_4, x_5, x_6
x_6	16	3	2	4	
z	0	-3	-5	-4	

			$\downarrow$		
		$-x_1$	$-x_4$	$-x_3$	
x_2	2.67	0.67	0.33	0	
$\leftarrow$ x_5	4.67	-1.33	-0.67	$\boxed{5}$	2nd basis: x_2, x_5, x_6
x_6	10.67	1.67	-0.67	4	
z	13.33	0.33	1.67	-4	

		$\downarrow$			
		$-x_1$	$-x_4$	$-x_5$	
x_2	2.67	0.67	0.33	0	
x_3	0.93	-0.27	-0.13	0.2	3rd basis: x_2, x_3, x_6
$\leftarrow$ x_6	6.93	$\boxed{2.73}$	-0.13	-0.8	
z	17.07	-0.73	1.13	0.8	

		$-x_6$	$-x_4$	$-x_5$	
x_2	0.98	-0.24	0.37	0.20	
x_3	1.61	0.10	-0.15	0.12	4th basis: x_2, x_3, x_1
x_1	2.54	0.37	-0.05	-0.29	
z	18.93	0.27	1.10	0.59	

With this, the final tableau has been obtained and with it the solution $x_1 = 2.54$, $x_2 = 0.98$, $x_3 = 1.61$, $x_4 = x_5 = x_6 = 0$, and $z = 18.93$.

1.3 Determination of a Feasible Initial Solution

For either the maximization or the minimization problem it is often not possible to exhibit a feasible initial solution immediately (as, for example, the trivial solution $x_1 = x_2 = \cdots = x_n = 0$). Moreover, it can happen that the system of constraints contains equations instead of inequalities.

In these two cases it is necessary to compute a first feasible initial solution with the help of special algorithms. Two such methods follow:

1.3.1 THE M-METHOD

This method can best be explained with the help of an example. Suppose the problem is to minimize the objective function

$$C = 3x_1 + 5x_2 + 4x_3 \tag{1.18}$$

subject to the constraints

$$
\begin{aligned}
2x_1 + 3x_2 - x_4 &= 8 \\
2x_2 + 5x_3 - x_5 &= 10 \\
3x_1 + 2x_2 + 4x_3 - x_6 &= 16 \\
x_i \geq 0 \quad (i = 1, 2, &\ldots, 6).
\end{aligned}
\tag{1.19}
$$

We may either assume that the three constraints in (1.19) were turned into equations by introduction of the slack variables x_4, x_5, and x_6, or that the system of constraints consisted of three equations from the very beginning. The M-method works equally well under both assumptions. In each equation another so-called artificial slack variable z_i is introduced, namely

$$
\begin{aligned}
2x_1 + 3x_2 - x_4 + z_1 &= 8 \\
2x_2 + 5x_3 - x_5 + z_2 &= 10 \\
3x_1 + 2x_2 + 4x_3 - x_6 + z_3 &= 16 \\
x_1, x_2, \ldots, x_6 &\geq 0 \\
z_1, z_2, z_3 &\geq 0.
\end{aligned}
\tag{1.20}
$$

For the extended system (1.20) it is always possible to find a feasible basic solution immediately by using the set of the artificial slack variables as a first basis and setting all other original as well as slack variables equal to zero. For system (1.20) we then have

$$z_1 = 8, \qquad z_2 = 10, \qquad z_3 = 16$$

and

$$x_1 = x_2 = \cdots = x_6 = 0.$$

Since the artificial slack variables have no meaning in the optimal solution, they necessarily must assume the value zero there. In order to eliminate the artificial slack variables from the basis as quickly as possible, the objective function (1.18) is modified by introduction of a so-called weight M, as follows:

$$C = 3x_1 + 5x_2 + 4x_3 + M(z_1 + z_2 + z_3). \qquad (1.21)$$

In the case of the minimization problem, the weight M in (1.21) is given a positive value which is large in comparison to the other coefficients of C, with the automatic result that during the iteration the artificial slack variables disappear from the basis. In the maximization problem the weight M has to have a correspondingly large negative value (cf. in this connection also Krelle and Künzi [87]). We can also set $M = 1$ and then minimize the expression $z_1 + z_2 + z_3$ until it is equal to zero.

One noticeable disadvantage of this method is the fact that the number of variables is in some cases very much enlarged. In the next section another relevant method is presented in which the number of variables does not increase.

1.3.2   The Multiphase Method[1]

This method will first be treated in general terms using the maximization problem as an example. The minimization problem can then be discussed analogously.

Maximize

$$B = \sum_{i=1}^{n} a_{0i} x_i$$

subject to the constraints

$$\sum_{i=1}^{n} a_{ji} x_i + x_{n+j} = a_{j0} \qquad (j = 1, \ldots, m)$$

$$x_i \geq 0 \qquad (i = 1, \ldots, m + n). \qquad (1.22)$$

[1] From Künzi [91].

Now suppose that the trivial initial solution $x_1 = x_2 = \cdots = x_n = 0$ representing the origin of the coordinate system, violates some of the constraints in the original system (1.1). Without loss of generality we can assume that these are the first r inequalities ($r \leq m$).

These r constraints are now dropped and the system of Eq. (1.22) is replaced by

$$\sum_{i=1}^{n} a_{ji}x_i + x_{n+j} = a_{j0} \qquad (j = r+1, \ldots, m). \qquad (1.23)$$

From among the disregarded constraints

$$\sum_{i=1}^{n} a_{ji}x_i + x_{n+j} = a_{j0} \qquad (j = 1, \ldots, r) \qquad (1.24)$$

any one is now singled out, for instance the first one with $j = 1$. This constraint is solved with respect to the slack variable and then used as an "artificial objective function" to be maximized subject to the constraints (1.23). This means that the new system to be investigated has the following form:

Maximize

$$x_{n+1} = \sum_{i=1}^{n} a_{1i}(-x_i) + a_{10}$$

subject to the constraints

$$\left.\begin{array}{c} \\ \sum_{i=1}^{n} a_{ji}x_i + x_{n+j} = a_{j0} \qquad (j = r+1, \ldots, m) \\ x_i \geq 0 \qquad (i = 1, \ldots, n, n+r+1, \ldots, n+m) \end{array}\right\} . \qquad (1.25)$$

In the new domain, in which the origin represents a feasible basic solution, we now iterate until the new objective function of (1.25), i.e., x_{n+1}, has become nonnegative. As soon as this condition ($x_{n+1} \geq 0$) is satisfied, the newly found vector, which shall be called z_1, has the property of satisfying the "dropped" constraint for $j = 1$, in addition to the previous constraints.

Following this first phase we investigate which constraints continue to be violated by the new vector z_1. It is certainly possible that the new point satisfies not only that one of the dropped constraints for $j = 1, 2, \ldots, r$ used as artificial objective function, but also others among them. Suppose that (perhaps after a suitable renumeration) the constraints $j = 2, \ldots, s$ are still violated at z_1.

This starts phase 2 where, precisely as in the first phase, the objective function

$$x_{n+2} = -a_{21}x_1 - a_{22}x_2 - \cdots - a_{2n}x_n + a_{20}$$

is optimized, subject to the constraints

$$\sum_{i=1}^{n} a_{ji}x_i + x_{n+j} = a_{j0} \qquad (j = 1, s + 1, \ldots, m)$$

$$x_i \geq 0 \quad (i = 1, \ldots, n, n + 1, n + s + 1, \ldots, n + m)$$

$$(1.26)$$

(Again we maximize until $x_{n+2} \geq 0$).

After finitely many such phases the obtained iteration point z_p satisfies all constraints and can be used as a starting point for the general simplex algorithm.

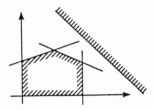

Fig. 3. An empty domain of feasible points for a linear optimization problem in the plane.

If the domain of feasible points is empty, as in Fig. 3 for example, the multiphase method shows this by the fact that

$$x_{n+j} \geq 0$$

cannot be obtained.

The multiphase method sketched here is characterized by three advantages:

(1)   There are no additional artificial variables.

(2)   Only the simplex algorithm is used for the determination of the first feasible basic solution.

(3)   During calculation on a computer the danger of floating point overflow is avoided.

Example 2.   Solve the problem given in 1.3, (1.18) and (1.19), in which

$$C = 3x_1 + 5x_2 + 4x_3$$

is to be minimized subject to the constraints

$$2x_1 + 3x_2 \qquad - x_4 = 8$$
$$2x_2 + 5x_3 - x_5 = 10$$
$$3x_1 + 2x_2 + 4x_3 - x_6 = 16$$
$$x_1, \ldots, x_6 \geq 0.$$

After introducing the artificial variables z_1, z_2, z_3, we obtain the starting tableau in which the negative objective function $-C = z$ is to be maximized:

			$-x_1$	$-x_2$	$-x_3$	$-x_4$	$-x_5$	$-x_6$
	z_1	8	2	3	0	-1	0	0
$\leftarrow$	z_2	10	0	2	$\boxed{5}$	0	-1	0
	z_3	16	3	2	4	0	0	-1
	z	$-34M$	$3 - 5M$	$5 - 7M$	$4 - 9M$	M	M	M

			$-x_1$	$-x_2$	$-z_2$	$-x_4$	$-x_5$	$-x_6$
	z_1	8	2	3	0	-1	0	0
	x_3	2	0	0.4	0.2	0	-0.2	0
$\leftarrow$	z_3	8	$\boxed{3}$	0.4	-0.8	0	0.8	-1
	z	$-16M - 8$	$3 - 5M$	$3.4 - 3M$	$0.8M - 0.8$	M	$0.8 - 0.8M$	M

			$-z_3$	$-x_2$	$-z_2$	$-x_4$	$-x_5$	$-x_6$
$\leftarrow$	z_1	2.67	-0.67	$\boxed{2.73}$	0.53	-1	-0.53	0.67
	x_3	2	0	0.4	0.2	0	-0.2	0
	x_1	2.67	0.33	0.13	-0.27	0	0.27	-0.33
	z	$-2.67M - 16$	$-1.67M$	$-2.73M + 3$	$-0.53M$	M	$0.53M$	$1 - 0.67M$

		$-z_3$	$-z_1$	$-z_2$	$-x_4$	$-x_5$	$-x_6$
x_2	0.98	-0.24	0.37	0.20	-0.37	-0.20	0.24
x_3	1.61	0.10	-0.15	0.12	0.15	-0.12	-0.10
x_1	2.54	0.37	-0.05	-0.29	0.05	0.24	-0.37
z	-18.93	-0.27	-1.10	-0.59	1.10	0.59	0.27

None of the artificial variables z_v is now in the basis any longer, and we can set $z_1 = z_2 = z_3 = 0$, i.e., delete the z_v-columns. As starting tableau of the original problem we therefore obtain:

		$-x_4$	$-x_5$	$-x_6$
x_2	0.98	−0.37	−0.20	0.24
x_3	1.61	0.15	−0.12	−0.10
x_1	2.54	0.05	0.29	−0.37
z	−18.93	1.10	0.59	0.27

And as can be seen right away, this tableau is already optimal and the solution is $x_1 = 2.54$, $x_2 = 0.98$, $x_3 = 1.61$, $C = 18.93$.

Example 3.   Using the multiphase method, find a feasible basic solution for Example 2.

Minimize:

$$C = 3x_1 + 5x_2 + 4x_3$$

subject to the constraints

$$
\begin{aligned}
-8 + 2x_1 + 3x_2 \quad\quad &\geq 0 \\
-10 \quad\quad + 2x_2 + 5x_3 &\geq 0 \\
-16 + 3x_1 + 2x_2 + 4x_3 &\geq 0 \\
x_1, x_2, x_3 &\geq 0.
\end{aligned}
$$

This problem is treated as a maximization problem with the objective function

$$B = -3x_1 - 5x_2 - 4x_3.$$

The origin of the coordinate system $x_1 = x_2 = x_3 = 0$ violates the constraints. We select the third constraint and use it as auxiliary objective function, obtaining as starting tableau:[1]

			x_1	x_2	x_3
	x_4	−8	2	3	0
	x_5	−10	0	2	5
←	x_6	−16	3	2	[4]
	z	0	−3	−5	−4

[1] Since this is not a feasible tableau, the notation is changed by placing in the top row not the negative variables as was done earlier, but the positive ones which are outside the basis.

The maximal coefficient of the auxiliary objective function determines the pivot row. We now look for the minimal quotient among all the satisfied constraints and, if such a quotient does not exist (as in the first step of our example), the auxiliary objective function itself can be used as pivot row. (Of course the condition for this must be that the maximal element of the auxiliary objective function is strictly positive. If this maximal element were to be nonpositive, no feasible basis would exist.)

Following the first exchange step we obtain the tableau

		x_1	x_2	x_6
$\leftarrow$ x_4	-8	2	$\boxed{3}$	0
x_5	10	$-\frac{15}{4}$	$-\frac{1}{2}$	$\frac{5}{4}$
x_3	4	$-\frac{3}{4}$	$-\frac{1}{2}$	$\frac{1}{4}$
z	-16	0	-3	-1

After this first step the last two constraints are satisfied. Now the first constraint has to be used as auxiliary objective function, and in this case it is again practical to use the auxiliary objective function as pivot row.

Following is the tableau after the second step:

		x_1	x_4	x_6
x_2	$\frac{8}{3}$	$-\frac{2}{3}$	$\frac{1}{3}$	0
x_5	$\frac{26}{3}$	$-\frac{11}{12}$	$-\frac{1}{6}$	$\frac{5}{4}$
x_3	$\frac{8}{3}$	$-\frac{5}{12}$	$-\frac{1}{6}$	$\frac{1}{4}$
z	-24	2	-1	-1

and with it a feasible basic solution has been found.

In order to arrive at an optimal tableau we continue the calculation with the simplex method. One further step is necessary.

1.4 The Fundamental Theorem and the Simplex Criterion of the Linear Optimization Theory

1.4.1 THE FUNDAMENTAL THEOREM

In this section we return once again to problem formulation (1.2) and assume throughout that the vector $a_{.0}$ cannot be represented as a positive linear combination of less than m columns of the matrix A, i.e., that degeneracy is excluded (cf. 1.5). The problem is to find a vector

$$x^{0T} = (x_1^{\,0}, \ldots, x_{n+m}^0)$$

representing an optimal feasible solution of (1.2).

If we denote the columns of the matrix A by A_i, and if x^0 exists, we have

$$a_0^T x^0 = \max a_0^T x$$

subject to the constraints

$$\left.\begin{array}{c} \displaystyle\sum_{i=1}^{n+m} x_i A_i = a_{.0} \\[2mm] x_i \geq 0 \qquad (i = 1, \ldots, n+m). \end{array}\right\} \qquad (1.28)$$

We shall now prove the fundamental theorem mentioned in 1.1.

Fundamental Theorem.   *If the optimal feasible vector x^0 of* (1.28) *exists, then a feasible basic solution also exists for which the objective function has the same optimal value.*

PROOF.   Without loss of generality, we can assume that the first k components of x^0 are strictly positive ($k \leq n + m$), while the remaining ones are zero. Suppose r is the maximal number of linearly independent columns among the first k columns A_i of the matrix A. By suitable renumeration we can then assure that the first r columns are linearly independent. Suppose that x^0 is not a basic solution, i.e., that $k > r$; we will show that an optimal feasible solution can be constructed from x^0 which has at most $k - 1$ nonzero components. Repetition of this process then leads to an optimal feasible basic solution after a finite number of steps.

We have

$$\sum_{i=1}^k x_i^{\,0} A_i = \sum_{i=1}^r x_i^{\,0} A_i + \sum_{i=r+1}^k x_i^{\,0} A_i = a_{.0} \qquad (1.29)$$

where

$$z = a_0^T x^0 = \sum_{i=1}^k a_{0i} x_i^{\,0} = z^0.$$

Since r is the maximal number of linearly independent vectors among the A_i, $i = 1, \ldots, k$, and since the vectors $A_1, \ldots, A_r$ are linearly independent, coefficients λ_i, $i = 1, \ldots, r$ clearly exist such that

$$A_k = \sum_{i=1}^r \lambda_i A_i. \qquad (1.30)$$

For the further discussion we associate an Δx_k with x_k, which, in view of (1.30), has to satisfy the relation

$$\Delta x_k A_k = \sum_{i=1}^{r} \Delta x_k \lambda_i A_i.$$ (1.31)

Then it follows from (1.29) and (1.31) that

$$\sum_{i=1}^{r} (x_i^0 - \Delta x_k \lambda_i) A_i + \sum_{i=r+1}^{k-1} x_i^0 A_i + (x_k + \Delta x_k) A_k = a_{.0}.$$ (1.32)

With this a new vector x^1 has been found which has components

$$x_i^1 = \begin{cases} x_i^0 - \Delta x_k \lambda_i & \text{for} \quad 1 \leq i \leq r \\ x_i^0 & \text{for} \quad r < i < k \\ x_i^0 + \Delta x_k & \text{for} \quad i = k \\ 0 & \text{for} \quad i > k \end{cases}$$

and for which the objective function has the value

$$z = \sum_{i=1}^{k} a_{0i} x_i^1 = \sum_{i=1}^{k} a_{0i} x_i^0 - \sum_{i=1}^{r} a_{0i} \Delta x_k \lambda_i + a_{0k} \Delta x_k$$

$$= z^0 + \Delta x_k \left(a_{0k} - \sum_{i=1}^{r} a_{0i} \lambda_i \right).$$ (1.33)

Here three cases are possible:

CASE 1

$$a_{0k} - \sum_{i=1}^{r} a_{0i} \lambda_i < 0.$$

Select Δx_k in such a way that

$$\Delta x_k = \max \left[\left\{ \frac{x_i^0}{\lambda_i} \middle| \lambda_i < 0, i = 1, \ldots, r \right\}; -x_k \right].$$ (1.34)

By assumption we have $x_i^0 > 0$ for $i = 1, \ldots, k$; certainly, therefore, because of (1.34),

$$\Delta x_k < 0$$

with which it follows from (1.33) that

$$z > z^0.$$

But this is a contradiction, since by assumption x^0 was optimal.

CASE 2

$$a_{0k} - \sum_{i=1}^{r} a_{0i}\lambda_i > 0.$$

It is up to the reader to work out a contradiction to this assumption in a manner similar to case 1.

CASE 3

$$a_{0k} - \sum_{i=1}^{r} a_{0i}\lambda_i = 0.$$

Analogously to case 1 the Δx_k are determined by (1.34); now, however, the value of the objective function remains unchanged while at least one of the first k components of the feasible vectors x^1 has to be zero. Therefore x^1 has only at most $k - 1$ nonzero components, and this completes the proof of the fundamental theorem.

It should be mentioned that in this proof the assumption excluding degeneracy was not used; the fundamental theorem is therefore valid in all cases.

1.4.2 THE SIMPLEX CRITERION

Under exclusion of degeneracies the simplex method shall now be discussed more fully. As we know, we start with a feasible basic solution of the problem (1.2), i.e., with a solution x with exactly m strictly positive components. Once again we will assume that these are the first m components. Accordingly, using the notation already introduced in (1.28) we have:

$$\left. \begin{array}{ll} \sum_{i=1}^{m} x_i A_i = a_{.0} & \\[2mm] x_i > 0 & (i = 1, \ldots, m). \\[2mm] x_i = 0 & (i = m + 1, \ldots, m + n). \end{array} \right\} \qquad (1.35)$$

Since x is a basic solution the columns $A_1, \ldots, A_m$ are linearly independent and therefore form a basis of the m dimensional vector space. In the third step of the simplex method, a variable x_e (e $> m$) is determined which is to enter the new basic solution and, more precisely (since degeneracy is excluded), $x_e > 0$ is chosen in such a way that exactly one variable equals zero, namely, the variable x_f (f $\leq m$) (determined in step 4) while all other variables

$$x_1, \ldots, x_{f-1}, \qquad x_{f+1}, \ldots, x_m$$

remain positive. Since $A_1, \ldots, A_m$ forms a basis of the m dimensional vector space, the column A_e belonging to x_e can be represented by

$$A_e = \sum_{i=1}^{m} \lambda_i A_i. \tag{1.36}$$

Hence,

$$x_e A_e = \sum_{i=1}^{m} x_e \lambda_i A_i,$$

and from (1.35) it follows that

$$\sum_{i=1}^{m} (x_i - x_e \lambda_i) A_i + x_e A_e = a_{.0}$$

with

$$\left.\begin{array}{l} x_e = \min_{1 \leq i \leq m}\left\{\left.\dfrac{x_i}{\lambda_i}\right| \lambda_i > 0\right\} = \dfrac{x_f}{\lambda_f} > 0 \\[2ex] x_i - x_e \lambda_i > 0 \qquad (i = 1, \ldots, f-1, f+1, \ldots, m) \\[1ex] x_f - x_e \lambda_f = 0. \end{array}\right\} \tag{1.37}$$

and

If we now define the vector x^1 by

$$x_i^1 = \begin{cases} x_i - x_e \lambda_i & (i = 1, \ldots, f-1, f+1, \ldots, m) \\ x_i, & (i = e) \\ 0, & (\text{otherwise}) \end{cases}$$

then x^1 is again a feasible vector. At this point it should be noted that when x_e, as defined above, does not exist (i.e., if $\lambda_i \leq 0$ for all i), no finite solution of the problem exists. In fact, in that case x_e and with it (according to step 3, condition 2), also the objective function can become arbitrarily large without violation of any sign restriction.

We will now show that the columns of the matrix A belonging to the positive components of x^1, i.e., $A_1, \ldots, A_{f-1}, A_{f+1}, \ldots, A_m, A_e$, are again linearly independent. The proof is indirect; in other words, we assume that a linear combination of these columns

$$\mu_1 A_1 + \cdots + \mu_{f-1} A_{f-1} + \mu_{f+1} A_{f+1} + \cdots + \mu_m A_m + \mu_e A_e$$

$$= \sum_{\substack{i=1 \\ i \neq f}}^{m} \mu_i A_i + \mu_e A_e = 0 \tag{1.38}$$

exists in which not all coefficients μ_i equal zero. Clearly, $\mu_e \neq 0$ because otherwise $A_1, \ldots, A_m$ would be linearly independent. Upon substitution of (1.36) it follows from (1.38) that

$$\sum_{\substack{i=1 \\ i \neq f}}^{m} \mu_i A_i + \sum_{i=1}^{m} \mu_e \lambda_i A_i = \sum_{\substack{i=1 \\ i \neq f}}^{m} (\mu_i + \mu_e \lambda_i) A_i + \mu_e \lambda_f A_f = 0. \qquad (1.39)$$

It follows from (1.37) that $\lambda_f > 0$; hence, since $\mu_e \neq 0$ we have $\mu_e \lambda_f \neq 0$ and (1.39) implies that the columns $A_1, \ldots, A_m$ are linearly independent, contradictory to our assumption. Therefore, the columns $A_1, \ldots, A_{f-1}$, $A_{f+1}, \ldots, A_m, A_e$ must be linearly independent. With this we have proved that (when degeneracy is excluded) each iteration step of the simplex method again provides a feasible basic solution with exactly m strictly positive components.

Let x be such a basic solution obtained after the νth iteration step. After suitable renumeration we can assure that exactly the first m components belong to the basis. Now define

$$_1a_0^T. = (a_{01}, \ldots, a_{0m}); \qquad _2a_0^T. = (a_{0,m+1}, \ldots, a_{0,m+n});$$

$$_1A = (A_1, \ldots, A_m); \qquad _2A = (A_{m+1}, \ldots, A_{m+n});$$

$$_1x^T = (x_1, \ldots, x_m); \qquad _2x^T = (x_{m+1}, \ldots, x_{m+n});$$

then

$$\left. \begin{array}{c} _1A \cdot {_1x} + {_2A} \cdot {_2x} = a_{.0} \\[2mm] _1x > 0, \qquad _2x = 0 \end{array} \right\} \qquad (1.40)$$

and the objective function has the value

$$z = {_1a_0^T.}\,{_1x} + {_2a_0^T.}\,{_2x}. \qquad (1.41)$$

As was just shown, the matrix $_1A$ is not singular; therefore, it follows from (1.40) that

$$_1x = {_1A^{-1}}a_{.0} - {_1A^{-1}}{_2A}\,{_2x}.$$

With this (1.41) changes to:

$$\begin{aligned} z &= {_1a_0^T.}{_1A^{-1}}a_{.0} - {_1a_0^T.}{_1A^{-1}}{_2A_2}x + {_2a_0^T.}{_2x} \\ &= {_1a_0^T.}{_1A^{-1}}a_{.0} + ({_2a_0^T.} - {_1a_0^T.}{_1A^{-1}}{_2A}){_2x}. \end{aligned} \qquad (1.42)$$

According to (1.40) we have $_2x = 0$ and hence

$$z = {_1a_0^T.}{_1A^{-1}}a_{.0} = a_{00}.$$

If the ith component $(i = 1, \ldots, n)$ of

$$_2a_{0.}^T - {_1a_{0.}^T}\,_1A^{-1}\,_2A$$

is positive, i.e., if

$$a_{0,m+1} - {_1a_{0.}^T}\,_1A^{-1}A_{m+i} > 0,$$

it follows from (1.42) that the value of the objective function can be increased by choosing $x_{m+i} > 0$ in such a way that neither the constraints nor the sign restrictions are violated. In order for x to be an optimal feasible solution it is necessary that

$$_2a_{0.}^T - {_1a_{0.}^T}\,_1A^{-1}\,_2A \leq 0. \tag{1.43}$$

The inequality (1.43) is called the *simplex criterion*. As (1.42) indicates, this criterion is evidently also a sufficient condition for the optimality of a basic solution $(_2x = 0)$.

A more thorough consideration of the vector on the left of (1.43) provides further insight into the simplex method. Suppose we have a feasible basis solution

$$x^T = (_1x^T, {_2x^T})$$

where $_1x > 0$ and $_2x = 0$, then

$$_1A\,_1x = a_{.0}. \tag{1.44}$$

But it is also possible that we have a feasible solution for which a component of $_2x$, for example x_{m+i}, is strictly positive, while all other components continue to be equal to zero. Then $_1x$ changes to $_1x - {_1\Delta x}$ whereby the m-vector $_1\Delta x$ is to be determined, such that

$$_1A(_1x - {_1\Delta x}) + x_{m+i}A_{m+i} = a_{.0}.$$

Hence, (1.44) implies that

$$_1A\,_1\Delta x = x_{m+i}A_{m+i}$$

or

$$_1\Delta x = x_{m+i}\,_1A^{-1}A_{m+i}.$$

Therefore, with respect to the equality constraints, the linear combination of the first m columns of A given by $_1A\,_1\Delta x$, is equivalent to the column given by $x_{m+i}A_{m+i}$. Accordingly, $_1\Delta x$ can be called an equivalent combination to x_{m+i}.

If we now consider the corresponding contributions to the value of the objective function, we see that x_{m+i} corresponds to the value

$$a_{0,m+i}x_{m+i}$$

while the value

$$_1a_{0.1}^T \Delta x = {}_1a_{0.1}^T A^{-1} A_{m+i} x_{m+i} \tag{1.45}$$

corresponds to the equivalent combination of x_{m+i}. Hence, the simplex criterion (1.43) states the following: The optimum is reached precisely then when the value of a unit of each nonbasic variable is not larger than the value of the corresponding equivalent combination. In this formulation the criterion also appears to be economically plausible.

1.4.3   THE PRICE VECTOR

The row vector $_1a_{0.1}^T A^{-1}$ appearing in (1.45) is often called the price vector π since it evidently plays an essential role in the value placed upon equivalent combinations. In other words, we set

$$\pi = {}_1a_{0.1}^T A^{-1}. \tag{1.46}$$

With (1.46), (1.43) then changes to

$$_2a_{0.}^T - \pi \, {}_2A \le 0. \tag{1.47}$$

In this form the simplex criterion will be used in the treatment of the decomposition principle (cf. 1.9).

1.4.4   THE OPTIMALITY CRITERION

We will now show that the optimality criterion given in the second step of the method is identical to the simplex criterion of (1.43). To this end we again consider a basic solution in which exactly the first m components are strictly positive. We then know that for this basic solution

$$_1x = {}_1A^{-1}a_{.0} - {}_1A^{-1}{}_2A \, {}_2x.$$

A comparison with tableau (1.17) shows that the coefficient matrix of this tableau is equal to the matrix $_1A^{-1}{}_2A$. Therefore, if $a_{i,m+j}$ is an element of tableau (1.17), then

$$_1A^{-1}{}_2A = \begin{pmatrix} a_{1,m+1} & a_{1,m+2} & \cdots & a_{1,m+n} \\ a_{2,m+1} & a_{2,m+2} & \cdots & a_{2,m+n} \\ \cdot & & & \\ \cdot & & & \\ \cdot & & & \\ a_{m,m+1} & a_{m,m+2} & \cdots & a_{m,m+n} \end{pmatrix}$$

And so, if $({}_1A^{-1}{}_2A)_i$ denotes the ith column of ${}_1A^{-1}{}_2A$, the ith component on the left side of (1.43) is

$$a_{0,m+i} - {}_1a_0^T({}_1A^{-1}{}_2A)_i = a_{0,m+i} - \sum_{j=1}^{m} a_{0j}a_{j,m+i} = -\alpha_{0,m+i},$$

as a comparison with (1.11) shows. Consequently, the simplex criterion of (1.43) assumes the form

$$-\alpha_{0,m+i} \leq 0 \qquad (i = 1, \ldots, n)$$

or

$$\alpha_{0,m+i} \geq 0 \qquad (i = 1, \ldots, n)$$

$$(1.48)$$

which is exactly the optimality criterion given in 1.2.[1]

1.4.5 PROOF OF THE FINITENESS OF THE SIMPLEX METHOD

In conclusion of this section we shall show that the simplex method leads to a solution in finitely many steps. We have assumed that only non-degenerate basic solutions occur. Moreover, we know on the basis of the simplex criterion or, equivalently, of condition 2 in the third step of the method, that the value of the objective function increases with each step of the iteration. Accordingly, the same basic solution cannot occur twice in the course of the method, since otherwise two different values of the objective function would belong to the same basic solution—which is obviously impossible. Since problem (1.2) has only finitely many feasible basic solutions, the optimum has to be reached after finitely many steps.

Throughout this section (with the exception of the fundamental theorem), degeneracies have been excluded. In the following section we will show that in the case of a degeneracy the problem can be modified in such a way that this degeneracy no longer occurs.

1.5 Degeneracies

In 1.1 we mentioned that in the two-dimensional case, i.e., for $n = 2$, we speak of degeneracy if, in the convex polygon determined by the constraints, more than two straight lines pass through one of the corner points. In general, we speak of degeneracy if in the convex polytope in R^n, formed by the system of constraints of (1.1), more than n hyperplanes pass through one of the corner points. Then, evidently, the feasible basic

---

[1] In the simplex tableau of Chapter 4 the π-values correspond to the α-coefficients belonging to the slack variables (cf. 1.11).

solution $_1x$ cannot consist entirely of positive elements

$$_1x_1, {}_1x_2, \ldots, {}_1x_m$$

but certain of these quantities have to be zero. From (1.44) it follows immediately that zero components occur in the vector $_1x$ when the vector $a_{.0}$ is linearly dependent upon some $(m - 1)$ columns of matrix $_1A$.

Further, it follows from the simplex algorithm given in 1.2 (step 4) that certain elements of the vector $_1x$ become zero if and only if, after the fourth step, the quotient

$$\frac{a_{j0}}{a_{je}} \quad \text{with} \quad a_{je} > 0$$

has the same minimal value for several j (the index e points to the entering variable, i.e., to the pivot column).

The following three statements are equivalent and each one permits ascertainment of the existence of degeneracy:

i.   More than n hyperplanes pass through the same corner point of the convex domain given by the constraints of (1.1).

ii. The vector $a_{.0}$ is linearly dependent upon some $m - 1$ columns of the matrix $_1A$ of (1.40).

iii. In the fourth step of 1.2 the value $\min_j \{a_{j0}/a_{je} \mid a_{je} > 0\}$ is assumed simultaneously for several j.

In the case of degeneracy the simplex criterion of 1.4 is no longer a necessary condition for the occurrence of a maximum.

If degeneracy occurs in the course of the iterative treatment of a problem, a supplementary method developed by Charnes [16] is useful for the removal of this difficulty, and hence for the solution of the apparent "deadlock" in the fourth step of the simplex algorithm.

The concept of Charnes is based on the fact that by a small displacement of the hyperplanes the coincidence of more than n such planes in the critical corner points is made impossible. These displacemen are made with the help of some vector ε and the modification

$$\begin{pmatrix} \bar{a}_{10} \\ \bar{a}_{20} \\ \cdot \\ \cdot \\ \cdot \\ \bar{a}_{m0} \end{pmatrix} = \begin{pmatrix} a_{10} \\ a_{20} \\ \cdot \\ \cdot \\ \cdot \\ a_{m0} \end{pmatrix} + \begin{pmatrix} a_{11} & a_{12} & \cdots & a_{1n} & 1 & 0 & \cdots & 0 \\ a_{21} & a_{22} & \cdots & a_{2n} & 0 & 1 & \cdots & 0 \\ \cdot \\ \cdot \\ \cdot \\ a_{m1} & a_{m2} & \cdots & a_{mn} & 0 & 0 & \cdots & 1 \end{pmatrix} \begin{pmatrix} \varepsilon^1 \\ \varepsilon^2 \\ \cdot \\ \cdot \\ \cdot \\ \varepsilon^{m+n} \end{pmatrix}$$

$$(1.49)$$

Here ε is an arbitrarily small but positive number. In (1.49) the column vector ε has as components the subsequent powers of ε from the first to the $(m + n)$th. (1.49) can be written in the matrix form

$$\bar{a}_{.0} = a_{.0} + A\varepsilon. \tag{1.50}$$

Both (1.49) and (1.50) show that the original vector $a_{.0}$ differs from the vector $\bar{a}_{.0}$ by a small displacement.

While, according to our above statements, the vector $a_{.0}$ is linearly dependent upon certain $m - 1$ columns of the matrix A, this can no longer be true for the new vector $\bar{a}_{.0}$, since the component $\bar{a}_{j0}$ of $a_{.0}$ contains the quantity ε^{n+j} not appearing in the other components $\bar{a}_{.j}$. Of course, the new problem will be arbitrarily close to the original one provided $\varepsilon > 0$ was sufficiently small.

We will now solve the programming problem for the new vector $\bar{a}_{.0}$ of (1.49). Due to the exclusion of degeneracy, $\bar{a}_{j0} = 0$ can no longer occur and hence the simplex criterion is valid. Therefore, we can select that variable x_i to enter the basis for which, after the third step in 1.2, the valuation coefficient α_{0i} is smallest. Then, provided that $\alpha_{0i} < 0$, we choose in line with the fourth step that variable x_v among those presently in the basis for which

$$\min_v \frac{\bar{a}_{v0}}{a_{ve}} = \min_v \left[\frac{a_{v0}}{a_{ve}} + \frac{{}_v a^T \varepsilon}{a_{ve}} \right]. \tag{1.51}$$

Here ${}_v a^T$ denotes the vth row of matrix A and the e refers to the pivot column. Even though a_{v0}/a_{ve} is simultaneously a minimum for several variables, $\bar{a}_{v0}/\bar{a}_{ve}$ will be a minimum only for one variable since, due to the forced linear independence, no variable in the basis can be equal to zero. If, for example, $a_{p0}/a_{pe} = a_{q0}/a_{qe}$ is simultaneously a minimum for two certain x-variables, then the terms

$$\frac{{}_r a^T \varepsilon}{a_{pe}} \quad \text{and} \quad \frac{{}_q a^T \varepsilon}{a_{qe}}$$

are to be considered for a decision. The smaller term determines the variable to be eliminated from the basis. We assumed that ε was an arbitrarily small positive quantity and that the vector contains the consecutive powers of ε. Therefore, the first term in the inner product determines the order of magnitude of all terms. Hence, we have to check whether a_{p1}/a_{pe} is larger or smaller than a_{q1}/a_{qe}, and, if both are equal, whether a_{p2}/a_{pe} is larger or smaller than a_{q2}/a_{qe}, and so on until equality ceases to exist and a decision has been reached.

With this criterion it is not possible to get into a cycle during the iteration, i.e., to return after finitely many steps once more to one of the former tableaus (cf. in this connection also Vajda [149, Chapter 5.3]). Hence, also in the case of degeneracy the simplex method terminates after finitely many steps.

1.6　Dual Linear Optimization Problems

Already in 1.1 we referred to the relationship between the maximization and the minimization problem at that time called duality.[1] In this section this duality concept will be explained in greater detail.

Two linear optimization problems are called dual if the following relation exists between them:

Maximization problem　　　　*Minimization problem*

Maximize the objective　　　　Minimize the objective
　　　function　　　　　　　　　　function

$$B = a_{01}x_1 + \cdots + a_{0n}x_n \qquad C = a_{10}w_1 + \cdots + a_{m0}w_m$$

subject to the $m + n$　　　　subject to the $n + m$
　　constraints　　　　　　　　　constraints

$$a_{11}x_1 + \cdots + a_{1n}x_n \leq a_{10} \qquad a_{11}w_1 + \cdots + a_{m1}w_m \geq a_{01}$$

$$a_{m1}x_1 + \cdots + a_{mn}x_n \leq a_{m0} \qquad a_{1n}w_1 + \cdots + a_{mn}w_m \geq a_{0n}$$
$$x_1, x_2, \ldots, x_n \geq 0. \qquad w_1, w_2, \ldots, w_m \geq 0.$$

$$(1.52)$$

It is frequently useful to combine the two constraint systems of (1.52) into the tableau

≥ 0	x_1	$x_2 \cdots x_n$	$(\leq)$
w_1	a_{11}	$a_{12} \cdots a_{1n}$	a_{10}
w_2	a_{21}	$a_{22} \cdots a_{2n}$	a_{20}
$\cdot$	$\cdot$		$\cdot$
$\cdot$	$\cdot$		$\cdot$
$\cdot$	$\cdot$		$\cdot$
w_m	a_{m1}	$a_{m2} \cdots a_{mn}$	a_{m0}
$\geq$	a_{01}	$a_{02} \cdots a_{0n}$	

$$(1.53)$$

---

[1] Frequently one also speaks of a primal- and dual problem.

Shorter still, the two systems of (1.52) can be presented in the matrix form

$$B = a_0^T. x \qquad C = w^T a_{.0}$$

$$Ax \leq a_{.0} \qquad A^T w \geq a_{0.} \qquad (1.54)$$

$$x \geq 0 \qquad w \geq 0.$$

Note that in (1.54) the matrices and vectors do not have the exact same meaning as in (1.2) and (1.7) since there the constraint system had already been transformed into a system of equations by the introduction of the slack variables. However, for the following considerations it will be necessary to use the forms (1.52) and (1.54).

The following two lemmas are of fundamental significance for the duality principle:

Lemma 1. *The inequality*

$$a_0^T. x \leq w^T a_{.0}$$

holds for any pair of feasible vectors x and w of problem (1.52).

PROOF. The relations

$$a_0^T. x \leq (w^T A)x = w^T(Ax) \leq w^T a_{.0}$$

are an immediate consequence of the constraints in (1.54).

Lemma 2. *If x^0 and w^0 are feasible and if*

$$a_0^T. x^0 = w^{0T} a_{.0}$$

then w^0 and x^0 are optimal.

PROOF. If $a_0^T. x^0 = w^{0T} a_{.0}$, it follows from Lemma 1 that

$$w^{0T} a_{.0} = a_0^T. x^0 \leq w^T a_{.0}$$

for all feasible w. Hence, w^0 is optimal. A similar consideration shows that x^0 too is optimal.

For the intended proof of the fundamental theorem of duality we need another lemma from the theory of linear inequalities.

Lemma 3. *The self-dual system*

$$Kz \geq 0, \qquad z \geq 0 \qquad (K^T = -K),$$

where K is a square skew-symmetric matrix, has solutions z_0 such that

$$Kz_0 + z_0 > 0.$$

The proof of Lemma 3 is fairly lengthy and requires some deeper knowledge of the theory of linear inequalities. The interested reader is hereby referred to the relevant original paper by Tucker [141].

Application of Lemma 3 to the skew-symmetric matrix

$$K = \begin{pmatrix} 0 & -A & a_{.0} \\ A^T & 0 & -a_{0.} \\ -a_{.0}^T & a_{0.}^T & 0 \end{pmatrix}$$

now yields the existence of a vector

$$z_0^T = (w^0, x^0, \lambda^0) \geq 0$$

for which the following inequalities are satisfied. Because of $Kz \geq 0$,

$$\begin{aligned} a_{.0}\lambda^0 &\geq Ax^0 \\ A^Tw^0 &\geq a_{0.}\lambda^0 \\ a_{0.}^Tx^0 &\geq a_{.0}^Tw^0 \end{aligned} \tag{1.55}$$

and because of $Kz_0 + z_0 > 0$,

$$\begin{aligned} w^0 + a_{.0}\lambda^0 &> Ax^0 \\ x^0 + A^Tw^0 &> a_{0.}\lambda^0 \\ \lambda^0 + a_{0.}^Tx^0 &> a_{.0}^Tw^0 . \end{aligned} \tag{1.56}$$

Two cases can now be distinguished for the scalar λ^0 introduced here.

Lemma 4. Let $\lambda^0 > 0$; then optimal feasible vectors x^0 and w^0 exist for the dual programs such that

$$a_{0.}^Tx^0 = a_{.0}^Tw^0$$
$$w^0 + a_{.0} > Ax^0$$
$$A^Tw^0 + x^0 > a_{0.} .$$

PROOF. Norm the vector z_0 in such a way that $\lambda_0 = 1$; this normalization does not affect the relations (1.55) and (1.56) (cf. also Tucker [141]). From the first two inequalities in (1.55) it then follows that x^0 and w^0 are feasible and the last inequality together with Lemma 1 yields:

$$a_{0.}^Tx^0 = a_{.0}^Tw^0 .$$

Lemma 2 now assures that x^0 and w^0 are optimal, and because of the first two relations of (1.56) we can use the normalized vectors x^0 and w^0 as the desired vectors x^0 and w^0.

Lemma 5. Let $\lambda^0 = 0$; then the following statements hold:

(a) Among the primal and the dual problem at least one has no feasible vector.

(b) If the maximization problem has a feasible vector, the set of its feasible vectors is unbounded and on this set $a_{0.}^T x$ is not bounded above. The corresponding result holds for the minimization problem.

(c) No problem has an optimal vector.

PROOF. Suppose x is a feasible vector for the maximization problem. Because of the nonnegativity of x it follows for $\lambda^0 = 0$ from the second relation of (1.55) that

$$x^T A^T w^0 \geq 0.$$

In view of the third relation of (1.56) and the feasibility of x this implies that

$$0 \leq x^T A^T w^0 \leq a_{.0}^T w^0 < a_{0.}^T x^0.$$

On the other hand, because of the first relation of (1.55), the assumption of the existence of a feasible vector w for the minimization problem would lead to the opposite inequality

$$0 \geq w^T A x^0 \geq a_{0.}^T x^0$$

and this proves (a).

For the proof of (b) we consider with the feasible vector x the ray

$$x + \mu x^0, \qquad \mu \geq 0 ;$$

then $x + \mu x^0 \geq 0$. With the aid of the first relation of (1.55) for $\lambda^0 = 0$ it then follows that

$$A(x + \mu x^0) \leq Ax \leq a_{.0}.$$

Consequently, the entire infinite ray consists of feasible vectors. This proves the first part of (b). Now, $a_{0.}^T(x + \mu x^0) = a_{0.}^T x + \mu a_{0.}^T x^0$ and, as already shown, $a_{0.}^T x^0 > 0$. This implies the second part of (b).

Finally, (c) is a direct consequence of (b).

Corollary. Either both the maximization and minimization problems possess optimal vectors or neither one does. In the first case the maximum and minimum are equal to each other and their common value is called the optimal value of the dual problems.

PROOF. If one of the problems has an optimal vector, it follows from Lemma 5(c) that $\lambda^0 > 0$ and from Lemma 4 that both problems have optimal vectors x^0 and w^0 and that the maximum $a_{0.}^T x^0$ is equal to the minimum $a_{.0}^T w^0$.

After these preparations we can now formulate the two fundamental theorems of linear optimization theory arising from the work of Dantzig, Gale, Kuhn, and Tucker.

I. The Duality Principle. *The feasible vector x^0 is optimal if and only if a feasible solution w^0 exists with*

$$a_{.0}^T w^0 = a_{0.}^T x^0 .$$

A feasible solution w^0 is optimal if and only if a feasible solution x^0 exists with

$$a_{0.}^T x^0 = a_{.0}^T w^0 .$$

PROOF. Lemma 2 shows that the condition is sufficient. In order to prove the necessity, assume that x^0 is optimal. According to Lemma 5(c), we then have $\lambda^0 > 0$. By Lemma 4 the minimization problem also has an optimal vector w^0 and it follows from the corollary that the maximum $a_{0.}^T x^0$ and the minimum $a_{.0}^T w^0$ are equal to each other. The second part of the principle is proved analogously.

II. The Existence Theorem. *A necessary and sufficient condition for one (and therefore both) dual problems to have optimal vectors is that both problems have feasible vectors.*

PROOF. The necessity is trivial. For the sufficiency of the condition, suppose that both problems have feasible vectors. Then by Lemma 5(a), $\lambda^0 > 0$ and by Lemma 4, both problems have optimal vectors.

1.7 The Dual Simplex Method

In view of the duality principle which we have just discussed it will be useful to switch to the dual method for the solution of certain problems. For this purpose we choose the representation already used at an earlier point, namely, (1.9) and (1.10):

Maximize:

$$z = a_{00} + \sum_{i=1}^{n} a_{0i}(-x_i) \tag{1.57}$$

subject to the constraints

$$x_{n+j} = a_{j0} + \sum_{i=1}^{n} a_{ji}(-x_i) \geq 0 \qquad (j = 1, \ldots, m) \tag{1.58}$$

$$x_i \geq 0 \qquad (i = 1, 2, \ldots, n).$$

As before, let us agree that for the initial solution the basis is formed by the variables, $x_{n+1}, \ldots, x_{n+m}$. In slight modification of the earlier representation of the tableau, we now write:

		$-x_1$	$-x_2$ $\cdots$ $-x_n$
z	a_{00}	a_{01}	a_{02} $\cdots$ a_{0n}
x_{n+1}	a_{10}	a_{11}	a_{12} $\cdots$ a_{1n}
x_{n+2}	a_{20}	a_{21}	a_{22} $\cdots$ a_{2n}
$\cdot$	$\cdot$	$\cdot$	$\cdot$
$\cdot$	$\cdot$	$\cdot$	$\cdot$
$\cdot$	$\cdot$	$\cdot$	$\cdot$
x_{n+m}	a_{m0}	a_{m1}	a_{m2} $\cdots$ a_{mn}

$$(1.59)$$

The following considerations are based on the same foundation as the theory of the ordinary simplex method and so we can restrict ourselves to a general description.

A tableau of the form (1.59) is called "primal-feasible" if all a_{j0} in the second lead column are nonnegative. Similarly, a tableau is said to be "dual-feasible" if all elements $a_{0i} \geq 0$. The tableau then represents a feasible solution of the dual problem.

By complete analogy, the dual simplex method transfers the well-known simplex steps to the dual problem.

If it is found that a problem is primal-feasible ($a_{j0} \geq 0$), the usual simplex method should be applied. If a problem is found to be primal-feasible we change over to the dual algorithm. It also follows from the duality principle that the optimal value has been reached if both the primal and the dual systems are feasible.

In order to start the iteration, one or the other system will of course have to be feasible. If this is not the case, we can find a first feasible solution with the help of the M-method or the multi-phase method. Once this solution has been obtained, the computational steps are as follows:

1. FOR THE PRIMAL-FEASIBLE TABLEAU

(a) Among the $a_{0i} < 0$ determine the largest one in modulus and take that variable newly into the basis which belongs to this absolutely largest a_{0i}, (let this be the variable x_e).

(b) For all positive a_{je} in the pivot column form the quotients a_{j0}/a_{je} and determine

$$\min_j \left\{ \frac{a_{j0}}{a_{je}} \right\}.$$

The row variable (say x_f) so determined has to be removed from the basis.

2. FOR THE DUAL-FEASIBLE TABLEAU

(a) Among all $a_{j0} < 0$ determine the one largest in modulus and remove the corresponding row variable (say x_h) from the basis.

(b) For all a_{hi} with $a_{hi} < 0$, form

$$\max_i \left\{ \frac{a_{0i}}{a_{hi}} \right\}$$

and choose the column variable determined in this way as the new basis variable. In both cases the value of a_{00} in the final tableau represents the optimal value of the objective function.

The case of degeneracy is treated in the same way as described in 1.5 for the ordinary simplex method.

In conclusion we mention that the dual simplex method discussed here is used centrally in the integer optimization algorithm by Gomory (see 1.11).

Example 4. Repeat the solution of Example 2 from 1.3 by using the dual simplex method. The dual initial tableau (1.59) then reads:

		$\downarrow$		
		$-x_1$	$-x_2$	$-x_3$
z	0	3	5	4
x_4	-8	-2	-3	0
x_5	-10	0	-2	-5
$\leftarrow$ x_6	-16	$\boxed{-3}$	-2	-4

Since all $a_{0i} \geq 0$, the above tableau is dually feasible, and for the first exchange step we determine among all $a_{j0} < 0$ the one with the largest modulus which is $x_h = x_6$. Further, we find

$$\max_i \left\{ \frac{a_{0i}}{a_{hi}} \right\} = -1 \quad \text{for} \quad i = 1$$

which means that x_1 has to be exchanged with x_6.

		$-x_6$	$-x_2$	$\downarrow$ $-x_3$
z	16	1	3	0
x_4	2.67	−0.67	−1.67	2.67
← x_5	−10	0	−2	−5
x_1	5.33	−0.33	0.67	1.33

Now x_5 has to be exchanged with x_3.

		$-x_6$	$\downarrow$ $-x_2$	$-x_5$
z	16	1	3	0
← x_4	−2.67	−0.67	−2.73	0.53
x_3	2	0	0.4	−0.2
x_1	2.67	−0.33	0.13	0.27

		$-x_6$	$-x_4$	$-x_5$
z	18.92	0.27	1.10	0.59
x_2	0.98	0.24	−0.37	−0.20
x_3	1.61	−0.10	0.15	−0.12
x_1	2.54	−0.37	0.05	0.29

This last tableau is primally and dually feasible and hence represents the optimal tableau. The solution is: $x_1 = 2.54$, $x_2 = 0.98$, $x_3 = 1.61$, $z = 18.92$.

1.8   The Revised Simplex Method

Dantzig *et al.* [39] have developed a modification of the simplex algorithm. It is natural to ask here whether it would not be possible to concentrate only on those entries in the tableau which are needed for the exchange steps. The revised simplex method—in the more current literature sometimes also called the inverse simplex method—concerns itself with this question.

Once again, nearly all considerations originate from the theory of the simplex method, so that here too we can restrict ourselves to a description of the new method. In line with formulation (1.2) we write the objective function to be maximized in the form

$$a_{0.}^T x + x_0 = 0 \qquad (1.60)$$

and the constraint system in the form

$$Ax = a_{.0}$$
$$x \geq 0. \qquad (1.61)$$

In (1.60) $-x_0$ denotes the quantity a_{00}. Now the objective function (1.60) and the constraint system (1.61) are combined as follows:

$$Bx = a_{.0} \qquad (1.62)$$

where

$$B = \left(\begin{array}{c|c|c} a_{0.}^T & 1 & 0^T \\ \hline A & 0 & I \end{array} \right) = \left(\begin{array}{c|c} a_{0.}^T & \\ \hline A & I \end{array} \right)$$

and

$$x^T = (x_1, \ldots, x_n \mid x_0 \mid x_{n+1}, \ldots, x_{n+m})$$
$$a_{.0}^T = (0 \mid a_{10}, a_{20}, \ldots, a_{m0}).$$

Furthermore, we have

$$x_1, x_2, \ldots, x_{n+m} \geq 0.$$

The new variable x_0 is allowed to be negative, but the other x_i are not.

This formulation has the advantage that all simplex calculations can be executed in one system, namely (1.62). Now the basis is exchanged repeatedly in such a way that in the case of the maximization problem x_0 decreases more and more while in the case of the minimization problem it grows increasingly larger.

It is easy to see that the method summarized below can be handled more effectively on an electronic computer than the ordinary simplex method.

The Algorithm for the Revised Simplex Method

STEP 1 (Determination of the Inverse Matrix B_i^{-1}). Remove from the matrix B all those columns belonging to the basic variables, including the column belonging to x_0. Form the inverse B_i^{-1} of this reduced $(m + 1) \times (m + 1)$ matrix B_i. The row vectors of B_i^{-1} are denoted by β_j, $j = 1, \ldots, m, m + 1$.

STEP 2 (Determination of the Variable x_e Entering the Basis).   Compute the quantities

$$\gamma_{0i} = \beta_1{}^T b_i \tag{1.63}$$

where β_1 is the first row of B_i^{-1} and b_i is the ith column of B, and the index i in (1.63) extends over all variables not in the basis. For the maximization problem the element x_e entering the basis is determined in such a way that

$$\max_i \gamma_{0i} = \beta_1{}^T b_e > 0.$$

If all

$$\gamma_{0i} \leq 0, \tag{1.64}$$

then the maximum has been reached.

For the minimization problem we use

$$\min_i \gamma_{0i} = \beta_1{}^T b_e < 0 \tag{1.65}$$

and if all $\gamma_{0i} \geq 0$, the minimum has been reached.

It is left up to the reader to prove that the criterion (1.64) is identical to the simplex criterion (1.43) given in 1.4, and that the quantity $\beta_1{}^T$ introduced above is given by

$$\beta_1{}^T = (1, -\pi) \tag{1.66}$$

where π is the price vector discussed in (1.46).[1]

STEP 3 (Determination of the Variable x_f Leaving the Basis).   In order to determine the exiting variable x_f we compute the quantities

$$y_{je} = \beta_j{}^T b_e \qquad (j = 2, \ldots, m+1).$$

For $y_{je} > 0$ the exiting variable x_f is then given by

$$\min_j \frac{a_{j0}}{y_{je}} = \frac{a_{f0}}{y_{fe}}.$$

Again, these exchange steps are analogous to those of the ordinary simplex algorithm.

For each step of the iteration we have to form a new inverse matrix B_i^{-1}. The objective will have been reached once all y_{0i} have become nonpositive in the case of the maximization problem and nonnegative in that of the minimization problem.

[1] For the proof compare, for example, Dantzig [33], Gass [62], and Krelle and Künzi [87].

We note that in the revised simplex algorithm only the inverse matrices have to be formed, in contrast to the general simplex method where the entire tableau has to be transformed every time. When an electronic computer is used this means the execution of fewer operations than in the ordinary algorithm, a reason why, as mentioned above, this method is particularly well-suited to automatic computation (cf. Wagner [153]). For this algorithm especially useful tableau arrangements can be constructed (cf. Krelle and Künzi [87] and Künzi and Tan [97]). A further extension of the revised simplex method can be found in the next section which discusses the decomposition principle.[1]

Example 5.   Repeat the solution of Example 1 in 1.3 by means of the revised simplex method. The matrix B and the column vector $a_{.0}$ then have the form:

$$B = \begin{bmatrix} x_1 & x_2 & x_3 & x_0 & x_4 & x_5 & x_6 \\ 3 & 5 & 4 & 1 & 0 & 0 & 0 \\ 2 & 3 & 0 & 0 & 1 & 0 & 0 \\ 0 & 2 & 5 & 0 & 0 & 1 & 0 \\ 3 & 2 & 4 & 0 & 0 & 0 & 1 \end{bmatrix}$$

$$a_{.0}^T = (0, 8, 10, 16).$$

Since $x_1 = x_2 = x_3 = 0$ are not contained in the first basis, we obtain as the first reduced matrix

$$B_1 = \begin{bmatrix} 1 & 0 & 0 & 0 \\ 0 & 1 & 0 & 0 \\ 0 & 0 & 1 & 0 \\ 0 & 0 & 0 & 1 \end{bmatrix}, \quad \text{and hence} \quad B_1^{-1} = B_1.$$

In order to determine x_e, we now compute

$$\max_i \gamma_{0i} = \beta_1^T b_2 = 5 > 0.$$

This in turn means that x_2 has to enter the basis.

For the determination of x_f we then calculate

$$\min_j \left\{ \frac{a_{j0}}{\beta_j^T b_2} \right\} = \frac{a_{20}}{\beta_2^T b_2} = \frac{8}{3},$$

[1] For other revised algorithms compare Künzi and Tan [97].

i.e., x_4 has to leave the basis. From this it follows that

$$B_2 = \begin{bmatrix} 1 & 5 & 0 & 0 \\ 0 & 3 & 0 & 0 \\ 0 & 2 & 1 & 0 \\ 0 & 2 & 0 & 1 \end{bmatrix} \quad \text{whence} \quad B_2^{-1} = \begin{bmatrix} 1 & -1.67 & 0 & 0 \\ 0 & 0.33 & 0 & 0 \\ 0 & -0.66 & 1 & 0 \\ 0 & -0.66 & 0 & 1 \end{bmatrix}.$$

In the next step x_5 is exchanged with x_3 and we obtain:

$$B_3^{-1} = \begin{bmatrix} 1 & -1.33 & -0.8 & 0 \\ 0 & 0.33 & 0 & 0 \\ 0 & -0.13 & 0.2 & 0 \\ 0 & -0.13 & -0.8 & 1 \end{bmatrix}.$$

The final tableau is obtained after an exchange of x_6 with x_1 :

$$B_4^{-1} = \begin{bmatrix} 1 & -1.1 & -0.59 & -0.27 \\ 0 & 0.37 & 0.19 & -0.24 \\ 0 & -0.15 & 0.12 & 0.10 \\ 0 & -0.05 & -0.29 & 0.37 \end{bmatrix}.$$

x_2, x_3, and x_1 form the basis. Since $\gamma_{0i} \leq 0$ for all i, the maximum has been reached. The values of the basic variables and of the objective function are found by calculating them in the same way as the quantities y_{je}, except that for b_e the vector $a_{.0}$ now has to be used:

$$z = \beta_1^T a_{.0} = 18.93 \qquad x_1 = \beta_4^T a_{.0} = 2.54$$
$$x_2 = \beta_2^T a_{.0} = 0.98 \qquad x_3 = \beta_3^T a_{.0} = 1.61$$

1.9 The Decomposition Algorithm

When linear optimization theory is used for the solution of certain practical problems from economy or industry, systems with very many constraints and variables frequently arise. In that case, the ordinary simplex algorithm generally fails to remain an effective method even when a large computer system is used. Dantzig and Wolfe [40] have developed an algorithm which in certain cases permits a decomposition into different partial problems and which then constructs the solution for the total problem from the partial solutions.

We modify the earlier notation slightly, writing the problem as follows: Minimize

$$L = \sum_{i=1}^{n} c_i^T x_i \tag{1.67}$$

subject to the constraints

$$\sum_{i=1}^{n} A_i x_i = b \tag{1.68}$$

$$B_i x_i = b_i \qquad (i = 1, \ldots, n) \tag{1.69}$$

$$x_i \geq 0.$$

In this formulation the following notation is used:

A_i is an $(m \times n_i)$-matrix B_i is an $(m_i \times n_i)$-matrix
c_i is an n_i-vector x_i is an n_i-vector
b is an m-vector b_i is an m_i-vector

Schematically, the problem here formulated can be represented by the following tableau:

$$\begin{array}{c}
 \\
\end{array}
\begin{array}{c c c c c c}
 & x_1 & x_2 & & x_n & \\
 & n_1 & n_2 & & n_n & \\
1 \left\{ \vphantom{\Big|} \right. & \boxed{c_1^T} & \boxed{c_2^T} & \cdots & \boxed{c_n^T} & \\
m \left\{ \vphantom{\Big|} \right. & \boxed{A_1} & A_2 & \cdots & \boxed{A_n} & b \\
m_1 \left\{ \vphantom{\Big|} \right. & \boxed{B_1} & & & & b_1 \\
m_2 \left\{ \vphantom{\Big|} \right. & & \boxed{B_2} & & & b_2 \\
\vdots & \vdots & & \ddots & & \vdots \\
m_n \left\{ \vphantom{\Big|} \right. & & & & \boxed{B_n} & b_n \\
\end{array} \tag{1.70}$$

For the further discussion, let us define the convex domains S_i by

$$S_i = \{ x_i \mid x_i \geq 0;\ B_i x_i = b_i \}. \tag{1.71}$$

At first, suppose that for all i these domains S_i are bounded. In addition, we are also interested in the set of all corner points

$$W_i = \{x_{i1}, x_{i2}, \ldots, x_{ik_i}\}$$

of the domains S_i. Now introduce the new quantities

$$P_{ik} = A_i x_{ik} \qquad (k = 1, \ldots, k_i, \quad i = 1, \ldots, n) \qquad (1.72)$$

$$c_{ik} = c_i^T x_{ik}. \qquad (1.73)$$

The principal concept of the decomposition algorithm now consists in constructing from the original problem (1.67)–(1.69) an equivalent formulation by using the quantities (1.72) and (1.73) as well as the new variables

$$z_{ik} \qquad (i = 1, \ldots, n, \quad k = 1, \ldots, k_i).$$

More precisely, minimize the objective function

$$\sum_{i=1}^{n} \sum_{k=1}^{k_i} c_{ik} z_{ik} \qquad (1.74)$$

subject to the constraints

$$\sum_{i=1}^{n} \sum_{k=1}^{k_i} P_{ik} z_{ik} = b \qquad (1.75)$$

as well as

$$\sum_{k=1}^{k_i} z_{ik} = 1 \qquad \text{for all} \quad i \qquad (1.76)$$

$$z_{ik} \geq 0 \qquad \text{for all} \quad i \quad \text{and} \quad k. \qquad (1.77)$$

This new formulation (1.74)–(1.76) can again be written in the tableau form

The row price-vector $(\boldsymbol{\pi}, \bar{\boldsymbol{\pi}})$ corresponds to the expression (1.46) and will be discussed further in (1.81). The connection between the two formulations (1.67)–(1.69) and (1.74)–(1.76) is contained in the following

Theorem.   *Let the quantities z_{ik} be the solution of* (1.75)–(1.77). *Then the vectors*

$$\boldsymbol{x}_i = \sum_{k=1}^{k_i} \boldsymbol{x}_{ik} z_{ik} \qquad (i = 1, \ldots, n)$$

form the solution of (1.67)–(1.69).

PROOF.   Let us denote problem (1.67)–(1.69) by I and problem (1.74)–(1.77) by II; then we need to show only that the minima of the objective functions C_1 and C_{11} are the same and that each feasible point of I corresponds uniquely to a feasible point of II.

(a) Let $\overset{\circ}{\boldsymbol{x}}_i$ be the solution of I. Then, because of (1.68) and (1.69), we have $\overset{\circ}{\boldsymbol{x}}_i \in S_i$. Hence, $\overset{\circ}{\boldsymbol{x}}_i$ can be represented by

$$\sum_{k=1}^{k_i} z_{ik} \boldsymbol{x}_{ik} \qquad \text{with} \qquad \sum_{k=1}^{k_i} z_{ik} = 1, \qquad z_{ik} \geq 0$$

and (1.68) changes to

$$\sum_{i=1}^{n} A_i \overset{\circ}{\boldsymbol{x}}_i = \boldsymbol{b} = \sum_{i=1}^{n} A_i \sum_{k=1}^{k_i} z_{ik} \boldsymbol{x}_{ik} = \sum_{i=1}^{n} \sum_{k=1}^{k_i} z_{ik} \boldsymbol{P}_{ik}.$$

Thus the constraints of II are satisfied and we have

$$C_{\mathrm{I}} = \sum_{i=1}^{n} \boldsymbol{c}_i^{T} \overset{\circ}{\boldsymbol{x}}_i = \sum_{i=1}^{n} \boldsymbol{c}_i^{T} \sum_{k=1}^{k_i} z_{ik} \boldsymbol{x}_{ik} = \sum_{i=1}^{n} \sum_{k=1}^{k_i} c_{ik} z_{ik} \geq C_{\mathrm{II}}. \qquad (1.79)$$

(b) Conversely, it is as easy to show that the optimal point of II corresponds to a feasible point of I and that therefore

$$C_{\mathrm{II}} \geq C_{\mathrm{I}}. \qquad (1.80)$$

The statement now follows from (1.79) and (1.80).

Returning once more to the new problem II, we note that it generally has more unknowns but fewer constraints than I. Except for the sign restrictions $z_{ik} \geq 0$, II has altogether $m + n$ constraints while in I

$$m + \sum_{i=1}^{n} m_i$$

equations occur.

Let the price vector of problem II corresponding to (1.46) be denoted by $(\boldsymbol{\pi}, \bar{\boldsymbol{\pi}})$ (where $\boldsymbol{\pi}$ is an m-vector and $\bar{\boldsymbol{\pi}}$ an n-vector). According to the discussion in 1.4, the value (1.74) can be decreased (provided there are feasible solutions at all) as long as

$$c_{ik} - \pi P_{ik} - 1 \cdot \bar{\pi}_i < 0. \tag{1.81}$$

If (1.81) does not hold by any column, and if degeneracy is excluded, then the optimum has been reached (see also the simplex criterion in (1.43)).

Suppose that a feasible basic solution with $m + n$ variables has been given. The corresponding basic vectors in tableau (1.78) have the form

$$\{P_{ik}, 0, \ldots, 1, \ldots, 0\}.$$

Suppose, furthermore, that the price vector $(\pi, \bar{\pi})$ with its $m + n$ components is known. Then, by (1.46), this vector and the basic vectors satisfy the equation

$$\pi P_{ik} + \bar{\pi}_i = c_{ik}. \tag{1.82}$$

One complete step of the iteration now has the following form:

STEP 1. For each i determine the solution of the sub-program

$$(c_i^T - \pi A_i)x_i \to \min \tag{1.83}$$

relative to

$$B_i x_i = b_i, \qquad x_i \geq 0. \tag{1.84}$$

Let $\overset{\circ}{x}_i$ be the corner point of S_i, solving this program.

STEP 2. Among these $\overset{\circ}{x}_i$ $(i = 1, \ldots, n)$ select that $\overset{\circ}{x}_{i0}$ for which

$$\gamma = (c_{i0}^T - \pi A_{i0})\overset{\circ}{x}_{i0} - \bar{\pi}_{i0} = \min_i [(c_i^T - \pi A_i)\overset{\circ}{x}_i - \bar{\pi}_i] \tag{1.85}$$

STEP 3. If $\gamma < 0$, the vector

$$(A_{i0}\overset{\circ}{x}_{i0}; 0, \ldots, 1, \ldots, 0)$$

is introduced into the basis and the price vector $(\pi, \bar{\pi})$ is computed anew according to the rules of the revised simplex method. If $\gamma \geq 0$ the optimal point of problem II and hence also of problem I has been found, as can be seen immediately by using the simplex criterion after substituting (1.72) and (1.73) into (1.85).

The finiteness of the process follows from the finiteness of the simplex method if we can demonstrate that for the indicated procedure $\gamma < 0$ always holds as long as (1.81) is still possible. This can be done as follows:

More explicitly written, (1.81) has the form

$$(c_i^T - \pi A_i)x_{ik} - \bar{\pi}_i < 0$$

where x_{ik} is a corner point of S_i for some i. For this i, (1.83) and (1.84) yield a corner point $\overset{\circ}{x}_i$ of S_i, which minimizes

$$(c_i^T - \pi A_i) x_i$$

and hence also

$$(c_i^T - \pi A_i) x_i - \bar{\pi}_i$$

over S_i. Since by assumption relation (1.81) holds, we have

$$\gamma \leq (c_i^T - \pi A_i)\overset{\circ}{x}_i - \bar{\pi}_i \leq (c_i^T - \pi A_i)x_{ik} - \bar{\pi}_i < 0.$$

Up to now it has been assumed that the sets S_i are bounded. If this is not the case it can happen that

$$\min_i \, [(c_i^T - \pi A_i)x_i], \qquad x_i \in S_i$$

is not bounded. For this it is necessary and sufficient that there exists a vector y_i for which

$$y_i \geq 0, \qquad B_i y_i = 0, \qquad \text{and} \qquad (c_i^T - \pi A_i)y_i < 0$$

(cf. Charnes *et al.* [20], and Dantzig and Wolfe [40]). In this case the column

$$(A_i y_i, 0, 0, \ldots, 0) \qquad \text{and} \qquad c_i^T y_i$$

is added to the extremal problem. This means that the z_{ik} belonging to this column is not contained in the condition $\sum_{k=1}^{k_i} z_{ik} = 1$. Otherwise, everything proceeds as before.[1]

Example 6. With the help of the decomposition algorithm, maximize

$$B = 18 + x_1 + 8x_2 + \tfrac{1}{2}x_3 + \tfrac{3}{2}x_4$$

subject to the constraints

$$
\begin{aligned}
x_1 + 4x_2 + \tfrac{7}{2}x_3 + \tfrac{1}{2}x_4 &\leq 1 \\
2x_1 + 3x_2 &\leq 6 \\
5x_1 + x_2 &\leq 5 \\
3x_3 - x_4 &\leq 12 \\
-3x_3 + x_4 &\leq 0 \\
x_3 &\leq 4 \\
x_1, x_2, x_3, x_4 &\geq 0.
\end{aligned}
$$

Here we have the case $n = 2$, and the convex domains S_1 and S_2 are defined

[1] For additional investigations of the problem of decomposition see Künzi and Tan [97].

by the partial constraints as follows:

$$S_1 = \left\{ (x_1, x_2) \,\middle|\, x_1, x_2 \geq 0, \begin{array}{r} 2x_1 + 3x_2 \leq 6 \\ 5x_1 + x_2 \leq 5 \end{array} \right\}$$

$$S_2 = \left\{ (x_3, x_4) \,\middle|\, x_3, x_4 \geq 0, \begin{array}{r} 3x_3 - x_4 \leq 12 \\ -3x_3 + x_4 \leq 0 \\ x_3 \qquad \leq 4 \end{array} \right\}$$

This example differs from the normal form (1.67), (1.68), and (1.69) in that inequality signs occur. Since the sub-problems can quite easily be solved graphically, it is not necessary to introduce slack variables for them. However, a slack variable x_5 is used for the first inequality. The given problem then has the tableau form

B	x_1	x_2	x_3	x_4	x_5	
1	-1	-8	$-\frac{1}{2}$	$-\frac{3}{2}$		$= 18$
	1	4	$\frac{7}{2}$	$\frac{1}{2}$	1	$= 1$
	2	3				≤ 6
	5	1				≤ 5
			3	-1		≤ 12
			-3	1		≤ 0
			1	0		≤ 4

maximize B; $x_1, x_2, x_3, x_4, x_5 \geq 0$.

A first feasible basis of the given system is

$$x_1 = x_2 = x_3 = x_4 = 0.$$

Let us denote by z_{11} and z_{21} the variables belonging to the corner points of S_1 and S_2, using $(x_1 = 0, x_2 = 0)$ and $(x_3 = 0, x_4 = 0)$, respectively. Then we obtain as the reduced tableau (1.78) the following initial tableau (without the last column):

	B	x_5	z_{11}	z_{21}		z_{22}
B	1	0	0	0	18	-20
$\leftarrow$ x_5	0	1	0	0	1	$\boxed{20}$
z_{11}	0	0	1	0	1	0
z_{21}	0	0	0	1	1	1

With the price vector $(\pi, \bar{\pi}) = (0, 0, 0)$ we find for (1.83) and (1.84):

B_1	x_1	x_2	
1	-1	-8	$= 0$
2	3	≤ 6
5	1	≤ 5

maximize B_1; $x_1, x_2 \geq 0$.

B_2	x_3	x_4	
1	$-\frac{1}{2}$	$-\frac{3}{2}$	$= 0$
3	-1	≤ 12
-3	1	≤ 0
1	0	≤ 4

maximize B_2; $x_3, x_4 \geq 0$.

The solutions are: $\overset{\circ}{x}_1{}^T = (0, 2)$, $\overset{\circ}{B}_1 = +16$, $\overset{\circ}{x}_2{}^T = (4, 12)$, $\overset{\circ}{B}_2 = +20$, i.e., in accordance with (1.85) we have $\gamma = -20$.

Since $\gamma < 0$, the optimum has not yet been reached and the vector $x_{22}^T = (A_2\overset{\circ}{x}_2, 0, 1) = (20, 0, 1)$ has to be introduced into the basis.

Let z_{22} be the corresponding variable. The variable leaving the basis is calculated to be x_5. By the exchange rules of the revised simplex method we obtain as a new tableau

	B	x_5	z_{11}	z_{21}		z_{12}
B	1	1	0	0	19	-8
$\leftarrow$ z_{22}	0	$\frac{1}{20}$	0	0	$\frac{1}{20}$	$\boxed{\frac{8}{20}}$
z_{11}	0	0	1	0	1	1
z_{21}	0	$-\frac{1}{20}$	0	1	$\frac{19}{20}$	$-\frac{8}{20}$

The new price vector $(\pi, \bar{\pi}) = (-1, 0, 0)$ leads to the partial problems:

B_1	x_1	x_2	
1	0	-4	$= 0$
2	3	≤ 6
5	1	≤ 5

maximize B_1; $x_1, x_2 \geq 0$

B_2	x_3	x_4	
1	3	-1	$= 0$
3	-1	≤ 12
-3	1	≤ 0
1	0	≤ 4

maximize B_2; $x_3, x_4 \geq 0$

with the solutions $\overset{\circ}{x}_1{}^T = (0, 2)$, $\overset{\circ}{B}_1 = +8$, $\overset{\circ}{x}_2{}^T = (0, 0)$, $\overset{\circ}{B}_2 = 0$. Hence, $\gamma = -8 < 0$ and the vector $x_{12}^T = (A_1\overset{\circ}{x}_1, 1, 0) = (8, 1, 0)$ has to be introduced into the basis.

The exchange of $\dot{z}_{12}$ with z_{22} yields:

	B	x_5	z_{11}	z_{21}		
B	1	2	0	0	20	$\gamma = 0$
z_{12}	0	$\frac{1}{8}$	0	0	$\frac{1}{8}$	
z_{11}	0	$-\frac{1}{8}$	1	0	$\frac{7}{8}$	
z_{21}	0	0	0	1	1	

and with $(\boldsymbol{\pi}, \bar{\boldsymbol{\pi}}) = (-2, 0, 0)$ we again get the partial problems

B_1	x_1	x_2	
1	1	0	$= 0$
	2	3	≤ 6
	5	1	≤ 5

maximize B_1; $x_1, x_2 \geq 0$

B_2	x_3	x_4	
1	$\frac{13}{2}$	$-\frac{1}{2}$	$= 0$
	3	-1	≤ 12
	-3	1	≤ 0
	1	0	≤ 4

maximize B_2; $x_3, x_4 \geq 0$

This yields

$$\overset{\circ}{x}_1{}^T = (0, 0), \qquad \overset{\circ}{B}_1 = 0$$

$$\overset{\circ}{x}_2{}^T = (0, 0), \qquad \overset{\circ}{B}_2 = 0.$$

Because $\gamma = 0$, the optimum has been reached. We find $\overset{\circ}{z}_{11} = \frac{7}{8}$, $\overset{\circ}{z}_{12} = \frac{1}{8}$, $\overset{\circ}{z}_{21} = 1$, $\overset{\circ}{B} = 20$ and for the original variables

$$(\overset{\circ}{x}_1, \overset{\circ}{x}_2) = \tfrac{7}{8}(0, 0) + \tfrac{1}{8}(0, 2)$$

$$(\overset{\circ}{x}_3, \overset{\circ}{x}_4) = 1(0, 0)$$

i.e., $\overset{\circ}{x}_1 = 0$, $\overset{\circ}{x}_2 = \frac{1}{4}$, $\overset{\circ}{x}_3 = 0$, $\overset{\circ}{x}_4 = 0$.

1.10 The Duoplex Algorithm

1.10.1 INTRODUCTION

The fundamental idea of the duoplex method (cf. Künzi [93], Künzi and Tzschach [98], and Künzi and Tan [97]) is based on the observation that in numerous cases it is possible quickly to find a constraint for which the bounding hyperplane contains the desired optimal point. This particular hyperplane is then made the new coordinate plane and with the help of the modified simplex algorithm those corner points of the polytope are determined which are contained in the special hyperplane. This particular algorithm is characterized by the fact that in a great many practical cases it requires fewer iteration steps for the solution of linear optimization problems than the simplex method. However, there are cases when the optimal point is not contained in the special hyperplane determined in the first step. In such cases the duoplex method still leads to the solution but in general we can no longer say that the number of steps remains smaller than for the simplex algorithm.

In general it can be said that the application of the duoplex algorithm is advantageous in those instances when the linear optimization problem has a large number of constraints.

1.10.2   THE ALGORITHM

We begin with the usual formula:
Maximize

$$z = a_{00} + \sum_{i=1}^{n} a_{0i}(-x_i)$$

subject to

$$x_{n+j} = a_{j0} + \sum_{i=1}^{n} a_{ji}(-x_j) \geq 0 \qquad (j = 1, \ldots, m)$$

$$x_i \geq 0 \qquad (i = 1, \ldots, n). \qquad (1.86)$$

As a starting point it is advantageous to use the origin $x_1 = 0, \ldots, x_n = 0$ even if it is not feasible.

STEP 1.   Determine that hyperplane for which the angle between its normal and the gradient of the objective function is largest. (In the representation (1.86) the normal vector of the constraints is given by $n_j^T = (-a_{j1}, \ldots, -a_{jn})$ and is directed toward the interior of the polytope.)

In many cases, the hyperplane selected in this way is the carrier of the optimal point. This is always so for $n \leq 2$, provided the chosen hyperplane (straight line) belongs to the boundary and hence is not redundant.[1]   For $n > 2$ this result need not be true.

---

[1] A constraint is called redundant if it does not belong to the boundary of the feasible domain.

The hyperplane can be determined using the scalar product of the different normal vectors n_j with the gradient of the objective function:

$$\min_{1 \le j \le m} \left\{ \frac{\sum\limits_{i=1}^{n} a_{0i} a_{ji}}{\left(\sum\limits_{i=1}^{n} a_{0i}^2\right)^{1/2} \left(\sum\limits_{i=1}^{n} a_{ji}^2\right)^{1/2}} \right\}. \qquad (1.87)$$

Suppose that in (1.87) the minimum is assumed for $j = j_p$; it is then also assumed for the same index if (1.87) is replaced by

$$\min_{1 \le j \le m} \left\{ \text{sign} \left(\sum\limits_{i=1}^{n} a_{0i} a_{ji}\right) \frac{\left(\sum\limits_{i=1}^{n} a_{0i} a_{ji}\right)^2}{\sum\limits_{i=1}^{n} a_{ji}^2} \right\}. \qquad (1.88)$$

From a numerical viewpoint, formula (1.88) is preferable over (1.87) since no square roots occur in it.

For the first exchange step determine

$$\max_{1 \le i \le n} \{-a_{0i}\}. \qquad (1.89)$$

If this maximum is assumed for $i = i_p$, the variable belonging to the j_pth column is to be exchanged with the one belonging to the j_pth row. In other words, the coordinate plane $x_{i_p} = 0$ is replaced by $x_{n+j_p} = 0$.

This first step of the duplex method occurs only once. Following it, we begin in steps 2, 3, and 4 the actual iteration process. Special mention should be made of the fact that in the duplex method it is permitted to move outside of the feasible domain because the algorithm will always lead back into this domain after finitely many steps.

STEP 2. Suppose r steps of the iteration have already been executed. This will be indicated by placing upper indices on the constants. Determine

$$\max_{1 \le i \le n} \{-a_{0i}^{(r)}\} \qquad (1.90)$$

and suppose the maximum is assumed for $i = i_0$. If $-a_{0i_0}^{(r)} \le 0$ and $a_{0j}^{(r)} \ge 0$ for all $j = 1, 2, \ldots, m$, then the given problem (1.86) is solved. Otherwise proceed to step 3.

STEP 3. In this step we try to increase the value of the objective function subject to the condition that none of the constraints already satisfied are violated by the step. For this purpose we determine the quantity λ by[1]

$$\lambda = \min_{1 \le j \le m} \left\{ \left| \frac{a_{j0}^{(r)}}{a_{ji_0}^{(r)}} \right| \; a_{j0}^{(r)} \ge 0 \wedge a_{ji_0}^{(r)} > 0 \right\}. \tag{1.91}$$

If λ exists and is assumed for $j = j_0$, and if $-a_{0i_0}^{(r)} > 0$, then, in accordance with the simplex method, exchange the variable x_{n+j_0} with x_{i_0} and continue with step 2. If λ does not exist, replace the value for λ by ∞ and continue with step 4. If $-a_{0i_0}^{(r)} \le 0$ but not $a_{j0}^{(r)} \ge 0$ for all j, continue likewise with step 4.

STEP 4. The purpose of this step is to satisfy as many of the violated constraints as possible, subject to the condition not to violate a constraint which has been satisfied before. For this determine the quantity

$$\sigma = \max_{1 \le j \le m} \left\{ \left| \frac{a_{j0}^{(r)}}{a_{ji_0}^{(r)}} \right| \; a_{j0}^{(r)} < 0 \wedge a_{ji_0}^{(r)} < 0 \wedge \frac{a_{j0}^{(r)}}{a_{ji_0}} \le \lambda \right\}. \tag{1.92}$$

Here four cases have to be distinguished.

i. If σ exists and is assumed for $j = j_1$, execute the exchange of x_{i_0} with x_{n+j_i} and continue with step 2. (Note that with this step at least one, and generally several, of the violated constraints will become satisfied. However, the value of the objective function will deteriorate.)

ii. If for $a_{j0}^{(r)} < 0$ all $a_{j_0}^{(r)} < 0$, and if λ does not exist, then, in the case of $-a_{0i_0}^{(r)} > 0$, no finite solution exists (cf. in this connection Charnes et al. [20]).

iii. If σ does not exist while λ does, and if there is at least one

$$a_{ji_0}^{(r)} < 0 \quad \text{when} \quad a_{j0}^{(r)} < 0, \tag{1.93}$$

then exchange the variable x_{n+j_0} with x_{j_0} and continue with step 2.

iv. If in case of iii the condition (1.93) cannot be satisfied for at least one j, return to step 2 but now exclude $i = i_0$. If there is no i, such that either $-a_{0i}^{(r)} > 0$ and λ exist, or, if there is at least one j so that

$$a_{j0}^{(r)} < 0 \quad \text{and} \quad a_{ji}^{(r)} < 0,$$

[1] In formula (1.91) the symbol $\wedge$ stands for the logical "and."

then the constraints are incompatible. It is easily shown that in this case at least one row exists in the tableau with

$$a_{j0}^{(r)} < 0 \quad \text{and} \quad a_{ji}^{(r)} \geq 0 \quad \text{for all} \quad i = 1, \ldots, n.$$

The finiteness of the duoplex method follows directly from the finiteness of the simplex- and the multiphase method (cf. Künzi [93] and Künzi and Tan [97]).

Example 7. Solve Example 1 from 1.3 by means of the duoplex method. Maximize

$$B = 3x_1 + 5x_2 + 4x_3$$

subject to the constraints

$$x_4 = 8 - 2x_1 - 3x_2 \qquad \geq 0$$
$$x_5 = 10 \qquad - 2x_2 - 5x_3 \geq 0$$
$$x_6 = 16 - 3x_1 - 2x_2 - 4x_3 \geq 0$$
$$x_1 \geq 0, \quad x_2 \geq 0, \quad x_3 \geq 0.$$

As a result of the first step we find that $j_p = 3$ and $i_p = 2$, and we therefore exchange x_6 with x_2; the tableau then reads

		$-x_1$	$-x_6$	$\downarrow$ $-x_3$
$\leftarrow$ x_4	-16	-2.5	-1.5	$\boxed{-6}$
x_5	-6	-3	-1	1
x_2	8	1.5	0.5	2
z	40	4.5	2.5	6

Two constraints are violated. Therefore, turning to step 4, we determine that the variables x_3 and x_4 are to be exchanged:

		$\downarrow$ $-x_1$	$-x_6$	$-x_4$
x_3	2.67	0.42	0.25	-0.17
$\leftarrow$ x_5	-8.67	$\boxed{-3.42}$	-1.25	0.17
x_2	2.67	0.67	0	0.33
z	24	2	1	1

One constraint remains violated, and, by the second and fourth step, x_1 and x_5 are selected for the exchange.

		$-x_5$	$-x_6$	$-x_4$
x_3	1.61	0.12	0.10	−0.15
x_1	2.54	−0.3	0.37	−0.05
x_2	0.98	0.19	−0.24	0.37
z	18.93	0.58	0.27	1.1

Using step 2 we find that this last tableau is feasible and hence optimal. Thus, $x_1 = 2.54$, $x_2 = 0.98$, $x_3 = 1.61$, $z = 18.93$ in the optimal solution.

Example 8.   By means of the duoplex method maximize

$$B = 11x_1 + 10x_2$$

subject to the constraints

$$x_3 = \;\; 0.8 + 0.5x_1 - 1.3x_2 \geq 0$$
$$x_4 = 10.7 - \;\; 4x_1 - \;\;\;\; x_2 \geq 0$$
$$x_5 = 15.4 - \;\; 6x_1 - \;\;\;\; x_2 \geq 0$$
$$x_6 = 13.4 - \;\; 6x_1 + \;\;\;\; x_2 \geq 0$$
$$x_7 = \;\; 8.7 - \;\; 4x_1 + \;\;\;\; x_2 \geq 0$$
$$x_8 = 10 \;\;\; - \;\; 5x_1 + \;\; 3x_2 \geq 0$$
$$x_1 \geq 0, \;\; x_2 \geq 0.$$

As a result of the first step we determine that x_1 and x_4 are to be exchanged. This leads to

		$-x_4$	$-x_2$
← x_3	2.14	0.13	[1.43]
x_1	2.67	0.25	0.25
x_5	−0.65	−1.5	−0.5
x_6	−2.65	−1.5	−2.5
x_7	−2	−1	−2
x_8	−3.37	−0.13	−4.25
z	29.4	2.75	−7.25

Four constraints are violated. Proceeding to step 4 we find that x_3 and x_2 have to be exchanged, which in turn provides the tableau

		$-x_4$	$-x_3$
x_2	1.5	0.09	0.70
x_1	2.3	0.23	-1.75
x_5	0.1	-1.46	0.35
x_6	1.1	-1.28	1.75
x_7	1.0	-0.82	1.40
x_8	3.0	-0.88	2.98
z	40.3	3.39	5.09

With this, the final tableau has been reached, and the solution is $x_1 = 2.3$, $x_2 = 1.5$, $z = 40.3$.

1.11   Linear Integer Optimization

In many practical applications it is demanded that the optimum calculated for a system like (1.1) or (1.16) should be integer-valued in the x_i. In other words, we add to the formulation (1.16) of the problem the requirement that all x_i in the optimal solution have integer values.

In order to satisfy this condition, appropriate special algorithms are needed; and here we shall describe the relevant points of the algorithm developed by Gomory [65]. Gomory's basic assumption is that all coefficients a_{ij} in (1.16) are rational numbers.

If once again we use a two-dimensional picture as illustration, the problem is to find, instead of the "farthest corner" of the convex polyhedron, the "farthest grid point" (cf. Fig. 4).

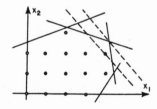

Fig. 4. Geometrical interpretation of the linear integer optimization problem.

We begin the algorithm by finding the noninteger optimum of the problem (here assumed to be the minimum) with the help of the simplex method. The resulting final tableau is analogous to (1.17), namely,

		$-x_{m+1}$	$-x_{m+2} \cdots -x_{m+n}$	
x_1	a_{10}	$a_{1,m+1}$	$a_{1,m+2} \cdots a_{1,m+n}$	
x_2	a_{20}	$a_{2,m+1}$	$a_{2,m+2} \cdots a_{2,m+n}$	(1.94)
.	.	.	. .	
.	.	.	. .	
.	.	.	. .	
x_m	a_{m0}	$a_{m,m+1}$	$a_{m,m+2} \cdots a_{m,m+n}$	

Now suppose that at least one of the values a_{j0} in tableau (1.94), for example a_{10}, is not yet an integer. Then every constant contained in the first row is split into its greatest integral part and the corresponding fractional part, i.e.,

$$a_{1i} = g_{1i} + f_{1i} ; \qquad a_{10} = g_{10} + f_{10} \tag{1.95}$$

where

$$0 \leq f_{1i} < 1 \qquad (i = m + 1, \ldots, m + n)$$
$$0 < f_{10} < 1 . \tag{1.96}$$

With (1.95) the first row of tableau (1.94) now changes into

$$g_{10} + f_{10} = x_1 + (g_{1,m+1} + f_{1,m+1})x_{m+1} + \cdots + (g_{1,m+n} + f_{1,m+n})x_{m+n} . \tag{1.97}$$

Equation (1.97) can be written as

$$f_{10} - [f_{1,m+1}x_{m+1} + \cdots + f_{1,m+n}x_{m+n}]$$
$$= x_1 + g_{1,m+1}x_{m+1} + \cdots + g_{1,m+n}x_{n+m} - g_{10} . \tag{1.98}$$

On the basis of the definitions $f_{1i} \geq 0$ and $x_{m+i} \geq 0$, the bracketed expression on the left of (1.98) is nonnegative for any feasible grid point. On the other hand, the right side of (1.98) corresponds by definition to an integer and can therefore not be larger than f_{10}. Since $f_{10} < 1$, it follows from this that

$$f_{1,m+1}x_{m+1} + \cdots + f_{1,m+n}x_{m+n} \geq f_{10} \tag{1.99}$$

or

$$x_{n+m+1} - f_{1,m+1}x_{m+1} - \cdots - f_{1,m+n}x_{m+n} = -f_{10} . \tag{1.100}$$

Here x_{n+m+1} is a new slack variable which by (1.98) has to be nonnegative and integer valued.

With the newly established relation (1.99) we have formed a new constraint which has to be satisfied by each (optimal or nonoptimal) nonnegative integer valued solution of the problem. On the other hand, if the new equation (1.100) is added to the tableau (1.94), we obtain for x_{n+m+1} the value $-f_{10}$, i.e., a negative, fractional value. We see that, through the introduction of the additional equation (1.100), the formerly optimal (noninteger) solution has dropped out of the feasible domain. In other words, the additional equation has reduced the original domain

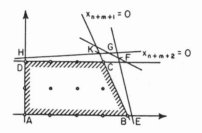

Fig. 5. Reduction of the feasible domain with help of the Gomory method.

without actually detaching an integer valued grid point. With the help of the dual simplex method we can now continue the calculation in such a way that x_{n+m+1} is no longer in the basis and therefore equal to 0.

If in this way an integer solution has been obtained, the objective has been reached. If fractional values occur again in the new solution, then, by the above procedure, new variables and equations are again introduced and the calculation is continued with the already mentioned dual simplex algorithm.

Generally speaking, the proposed Gomory method consists of a repeated reduction of the original domain by means of additional constraints, but in such a way that, as mentioned earlier, no grid points are cut away. Figure 5 illustrates this situation with a simple two-dimensional example.

Let the original convex domain $AEFGH$ be given, and suppose that the ordinary optimum is assumed at the point F (nongrid point). We are here particularly interested in the convex hull of the grid points (hatched domain). The algorithm described results in the following steps:

1. The simplex algorithm leads from A to F.

2.   Addition of the new inequality $x_{n+m+1} \geq 0$ reduces the original domain in such a way that the corner points E, F, and G no longer belong to the reduced domain.

3.   With the dual simplex method a new feasible solution is calculated by going from F via G to K.

4.   A new inequality $x_{n+m+2} \geq 0$ is added.

5.   With the help of a further iteration of the dually feasible tableau, we obtain the integer valued optimum at point C.

Each time a new equation and a new variable have to be introduced for the elimination of a fractional value. Accordingly, it is not immediately evident that the proposed method indeed does lead in finitely many steps to the result. However, in his cited paper Gomory has provided the finiteness proof, for which we here simply refer to the original publication [64].

Experience has also shown that in the case of many practical computations the number of iterative steps required can become very large, a fact which could lessen the practical usefulness of the Gomory method.

Furthermore, it is immediately evident that the additional equation can be formed in several different ways. In fact, if an equation is multiplied by a number λ, different fractions and hence different additional constraints are obtained. Altogether, we get Λ different cases where Λ is the greatest common denominator of the coefficients of an equation.

Example 9.   With the help of the Gomory algorithm maximize

$$B = -10x_1 + 111x_2$$

subject to the constraints

$$40 + x_1 - 10x_2 \geq 0$$

$$20 - x_1 - x_2 \geq 0$$

$$x_1, x_2 \geq 0 \, ; \qquad x_1, x_2 \quad \text{integer valued} \, .$$

Accordingly, the initial tableau has the form

		$-x_1$	$-x_2$
x_3	40	-1	$+10$
x_4	20	$+1$	$+1$
z	0	$+10$	-111

Basis: x_3, x_4

After two transformation steps we obtain for the noninteger valued optimum the tableau

		$-x_4$	$-x_3$
x_2	$\frac{60}{11}$	$\frac{1}{11}$	$\frac{1}{11}$
x_1	$\frac{160}{11}$	$\frac{10}{11}$	$-\frac{1}{11}$
z	460	1	11

Since not all a_{j0} $(j = 1, 2)$ are integer-valued, an additional constraint has to be introduced. For this we decompose

$$a_{20} = 5 + \tfrac{5}{11}$$
$$a_{24} = 0 + \tfrac{1}{11}$$
$$a_{23} = 0 + \tfrac{1}{11}$$

whence follows the additional constraint:

$$x_5 = -\tfrac{5}{11} - \tfrac{1}{11}x_4 - \tfrac{1}{11}x_3 \,;$$

thus the tableau reads:

$$\downarrow$$

		$-x_4$	$-x_3$	
x_2	$\frac{60}{11}$	$\frac{1}{11}$	$\frac{1}{11}$	
x_1	$\frac{160}{11}$	$\frac{10}{11}$	$-\frac{1}{11}$	Basis: x_2, x_1, x_5
$\leftarrow x_5$	$-\frac{5}{11}$	$\boxed{-\frac{1}{11}}$	$-\frac{1}{11}$	
z	460	1	11	

and after one further transformation we have:

		$-x_5$	$-x_3$	
x_2	5	1	0	
x_1	10	10	-1	Basis: x_2, x_1, x_4
x_4	5	-11	1	
	455	$+11$	10	

With this an integer-valued optimal tableau has been obtained and the solution is $x_1 = 10$, $x_2 = 5$, $x_4 = 5$, $x_3 = x_5 = 0$, $B = 455$.

2 NONLINEAR OPTIMIZATION

2.1 Convex Domains and Functions

Within the framework of this book it is not possible to treat the theory of nonlinear optimization in the same detail as was done for linear optimization in Chapter 1. Here we shall only be concerned with a brief survey of some of the principal ideas. For a more detailed presentation we refer to Boot [12] and Künzi and Krelle [95]. Because the following discussions are closely connected with the theory of convex domains and functions, we will begin with a brief survey of several definitions and properties of such domains and functions.

2.1.1 CONVEX DOMAINS

A point set M in the n dimensional Euclidean space R^n is called convex if with any two points z^1 and z^2 from M all points

$$\lambda z^1 + (1 - \lambda)z^2 \quad \text{with} \quad 0 \leq \lambda \leq 1$$

also belong to M. Geometrically, this means that a set is convex if, together with any two of its points, it also contains the connecting straight line segment between them.

If the set is closed it is said to be a convex domain. Such domains need not be bounded and they can be contained in a linear subspace of lower dimension than n (cf. Fig. 6).

The entire space R^n is convex and the empty set containing no elements can also be considered convex. Furthermore, the intersection of arbitrarily many convex sets is once again convex.

If a convex set M is not bounded, then for each point $z^0 \in M$ a ray $z^0 + \lambda t$ $(\lambda > 0)$, $(t \in R^n)$ can be constructed which is entirely contained

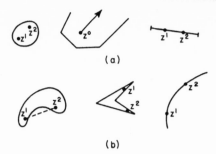

(a)

(b)

Fig. 6. (a) Convex and (b) nonconvex domains in the plane.

in M. If two convex sets M^1 and M^2 have at most boundary points in common, there exists a separating hyperplane $a^T z = b$ such that $a^T z^1 \leq b$ for all $z^1 \in M$ and $a^T z^2 \geq b$ for all $z^2 \in M^2$.

2.1.2 CONVEX FUNCTIONS

A function $F(x)$, $(x^T = (x_1, \ldots, x_n))$ on R^n is called convex on the convex domain M if for any two points x^1 and x^2 of M

$$F(\lambda x^1 + (1 - \lambda)x^2) \leq \lambda F(x^1) + (1 - \lambda)F(x^2)$$

holds for $0 < \lambda < 1$.

The function is called strictly convex if for $x^1 \neq x^2$ the symbol $\leq$ can be replaced by $<$. Along the straight line segment between x^1 and x^2 a convex function cannot assume a value larger than that of the linear function obtained by linear interpolation (Fig. 7).

$F(x)$ is called concave (strictly concave), if $-F(x)$ is convex (strictly convex). Hence a linear function is both convex and concave. Without entering into their proofs we note the following properties of convex (differentiable) functions:

i.
$$F(x^2) - F(x^1) \geq (x^2 - x^1)\left(\frac{\partial F}{\partial x}\right)_{x^1} \qquad \text{for all} \quad x^1, x^2.$$

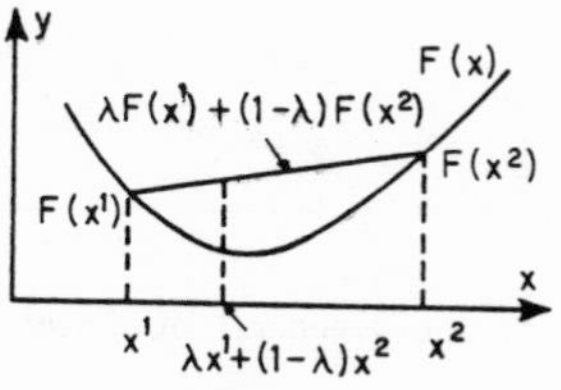

Fig. 7. Geometrical interpretation of a convex function in the plane.

ii.   The matrix $(\partial^2 F/\partial x_i\,\partial x_j)$ of the second derivatives is positive semi-definite for all x.

iii.   If C is a positive semi-definite matrix, the function

$$Q(x) = x^T C x$$

is convex; if C is strictly positive definite, then $Q(x)$ is strictly convex.

In conclusion it should be mentioned that for a convex function each local minimum of the convex domain M is a global minimum on this domain. The set of all its minima is once again convex. A strictly convex function has at best one local minimum.

2.2   General Nonlinear Optimization

If the linear optimization problem is extended in the most general way to the nonlinear case, i.e., in particular, if both the objective function and the constraints are assumed to be nonlinear, then we obtain the following formulation:

Find the minimum of the objective function

$$G(x), \qquad x^T = (x_1, \ldots, x_n)$$

subject to the constraints

$$g_j(x) \leq 0 \qquad (j = 1, \ldots, m),$$

$$x_i \geq 0 \qquad (i = 1, \ldots, n).$$

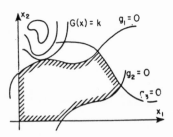

Fig. 8. Geometric interpretation of a general, nonlinear, optimization problem in the plane.

For the two dimensional case, Fig. 8 illustrates such an optimization problem. However, we are today still far from the solution of such a general optimization problem. Instead, we have to rely on more or less effective trial and error methods which generally lead only into the neighborhood of certain secondary optima. It must be considered a stroke of luck if the absolute optimum is attained. Still, good methods and formulations are available for a very important special case as described in 2.3.

2.3 Convex Optimization

Let $F(x)$ and $f_j(x), j = 1, \ldots, m$ be convex functions of the n variables $x_1, \ldots, x_n$. Under convex optimization we understand the problem of minimizing the function $F(x)$ subject to the constraints

$$f_j(x) \leq 0 \qquad (j = 1, \ldots, m)$$
$$x \geq 0. \tag{2.1}$$

In shorter form this means that

$$\min \{F(x) \,|\, x_i \geq 0 \ \text{ for } \ i = 1, \ldots, n, \ \text{ and } \ f_j(x) \leq 0 \ \text{ for } \ j = 1, \ldots, m\}$$

Figure 9 illustrates the situation in the two-dimensional case. As in the case of linear optimization, $F(x)$ is called the objective function. A point x satisfying the constraints $f_j(x) \leq 0$ in (2.1) is called a feasible point. The set R of all feasible points is the feasible domain. Because of the assumed convexity of the f_j, R is also convex. The program is called feasible if R is not empty. The hypersurfaces given by $f_j(x) = 0$ and $x_i = 0$

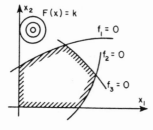

Fig. 9. Geometrical interpretation of a convex optimization problem in the plane.

are said to be boundary planes of the feasible domain provided they contain feasible points. A feasible program is called solvable if the objective function is bounded on R and actually assumes its minimum on R. If R is bounded and not empty, at least one solution exists.

2.4 The Kuhn–Tucker Conditions

The Kuhn–Tucker theorem is the central theorem for nonlinear optimization. It represents a generalization of the classical method of Lagrange multipliers for the determination of extrema under constraints, to include the case when these constraints not only contain equations but inequalities as well. More precisely, it provides necessary and sufficient conditions for a certain $\hat{x}$ to be a solution of problem (2.1) (cf. [89]).

These conditions utilize a so-called generalized Lagrange function Φ. We introduce m new variables $u_1, \ldots, u_m$, the Lagrange multipliers, and combine them into a vector u. Then Φ is a function of the $m + n$ variables (x, u) given by

$$\Phi(x, u) = F(x) + \sum_{j=1}^{m} u_j f_j(x).$$

The Kuhn–Tucker theorem now states:

A vector $\hat{x}$ is a solution of problem (2.1) if and only if a vector $\hat{u}$ exists such that

$$\hat{x} \geq 0, \qquad \hat{u} \geq 0 \tag{2.2}$$

and

$$\Phi(\hat{x}, u) \leq \Phi(x, u) \leq \Phi(x, \hat{u}) \tag{2.3}$$

for all

$$x \geq 0 \qquad \text{and} \qquad u \geq 0.$$

Hence, at $(\hat{x}, \hat{u})$, $\Phi(x, \hat{u})$ has to assume a global minimum with respect to x on the domain $x \geq 0$ as well as a global maximum with respect to u on the domain $u \geq 0$. This value can be called a nonnegative saddle point.

Therefore, we have associated with the minimization problem for $F(x)$ a saddle point problem (i.e., a minimax problem) for Φ in which the constraints consist only of sign restrictions. The x-part of the solution of the minimax problem represents a solution of the minimization problem. For the proof of this theorem we refer to the original paper [89] or the presentation by Künzi and Krelle [95].

If $F(x)$ and $f_j(x)$ are differentiable functions, the conditions (2.2) and

(2.3) are equivalent to

$$\left(\frac{\partial \Phi}{\partial x_i}\right)_{\hat{x},\hat{u}} \geq 0 \tag{2.4}$$

$$\hat{x}_i \left(\frac{\partial \Phi}{\partial x_i}\right)_{\hat{x},\hat{u}} = 0 \tag{2.5}$$

$$\hat{x}_i \geq 0 \tag{2.6}$$

$$\left(\frac{\partial \Phi}{\partial u_j}\right)_{\hat{x},\hat{u}} \leq 0 \tag{2.7}$$

$$\hat{u}_j \left(\frac{\partial \Phi}{\partial u_j}\right)_{\hat{x},\hat{u}} = 0 \tag{2.8}$$

$$\hat{u}_j \geq 0. \tag{2.9}$$

It can easily be shown that (2.2) and (2.3) imply (2.4)–(2.9), since for fixed, nonnegative $\hat{u}$, $\Phi(x, \hat{u})$ is a convex function of the x_i.

Now suppose that for $\hat{x}_i \geq 0$ the condition (2.4) or (2.5) is violated, i.e., that $\hat{x}_i > 0$ and $\partial\Phi(\hat{x}, \hat{u})/\partial x_i \neq 0$ or $\hat{x}_i \neq 0$ and $\partial\Phi(\hat{x}, \hat{u})/\partial x_i < 0$; then there exist points $x_i \geq 0$ for which Φ assumes a lower value than at $\hat{x}_i$ and this is a contradiction to the saddle point theorem. Correspondingly, it can be shown that conditions (2.4)–(2.9) imply (2.2) and (2.3).

The Kuhn–Tucker theorem permits certain variations in the formulation of problem (2.1).

(a)   It is possible that the constraint $x_i \geq 0$ is missing. In that case the three conditions (2.4), (2.5), and (2.6) are replaced by the one condition

$$\left(\frac{\partial \Phi}{\partial x_i}\right)_{\hat{x},\hat{u}} = 0. \tag{2.10}$$

(b)   If $f_j(x)$ is linear, a constraint of the form $f_j(x) = 0$ is feasible. From

$$\Phi(x, u) = F(x) + \sum_{j=1}^{m} u_j f_j(x),$$

we see that in this case (2.7), (2.8), and (2.9) simplify to

$$\left(\frac{\partial \Phi}{\partial u_j}\right)_{\hat{x},\hat{u}} = 0, \tag{2.11}$$

where the sign of u is no longer restricted.

(c)   If the problem has the form

$$\min \{F(x) \mid f_j(x) = 0\},$$

with linear constraints, only the two conditions (2.10) and (2.11) remain, and this is equivalent to the classical optimization theorem with constraints of differential calculus.

2.5   Quadratic Optimization

From an algorithmic viewpoint the special case of quadratic optimization deserves particular interest and for this reason will be discussed here in somewhat more detail.

The objective function is now assumed to have the form

$$\Phi(x) = a_0^T x + x^T C x$$

where C is a symmetric and positive definite or semidefinite matrix. As in Chapter I the constraints are assumed to be linear and given by

$$Ax \leq a_{.0}$$
$$x \geq 0.$$

We can distinguish three different types of problem formulations, namely:

$$\min \{\Phi(x) \mid Ax \leq a_{.0}, x \geq 0\} \tag{2.12}$$

$$\min \{\Phi(x) \mid Ax = a_{.0}, x \geq 0\} \tag{2.13}$$

$$\min \{\Phi(x) \mid Ax \leq a_{.0}\}. \tag{2.14}$$

In (2.13) the inequalities have been changed to equations by the introduction of slack variables, and in (2.14) the sign restrictions (if there were any) have been absorbed into the constraint system.

In the case of positive definite C, the figures (10a), (10b), and (10c) give a two-dimensional geometrical interpretation for the above problem formulation. Note the differences compared to linear optimization. In that case the objective function assumed its minimum in a corner of the convex polyhedron, while in the case of quadratic optimization with linear constraints, the minimum can be located at a corner (Fig. 10a), on an edge (10b), or even in the interior of the polyhedron (10c).

In other words, for a linear optimization problem (with no degeneracy), exactly n of the $m + n$ inequalities are satisfied as equations at the optimal point, while for the quadratic optimization this is the case for at most n of the $m + n$ inequalities.

The Lagrange function for the three problems (2.12), (2.13), and (2.14) now has the form

$$\Phi(x, u) = a_0^T x + x^T C x + u^T (Ax - a_{.0}). \tag{2.15}$$

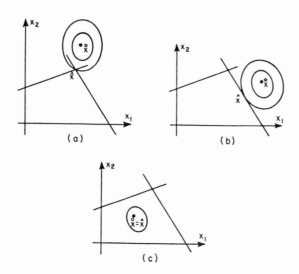

Fig. 10. Geometrical interpretation of quadratic optimization problems with linear constraints.

With the abbreviation

$$\frac{\partial \Phi}{\partial x} = v \quad \text{and} \quad -\frac{\partial \Phi}{\partial u} = y$$

we have

$$v = a_{0.} + 2Cx + A^T u = \frac{\partial \Phi}{\partial x} \qquad (2.16)$$

and

$$y = -Ax + a_{.0} = -\frac{\partial \Phi}{\partial u}. \qquad (2.17)$$

With these abbreviations the Kuhn–Tucker conditions formulated for problem (2.13) are as follows:

$$\left. \begin{array}{c} Ax = a_{.0} \\ 2Cx - v + A^T u = -a_{0.} \\ x \geq 0, \quad v \geq 0 \\ x^T v = 0 \end{array} \right\} \qquad (2.18)$$

2.6   Duality in the Case of Quadratic Optimization

The duality concept is less important in the nonlinear case than in the linear one since the symmetry between the two problems is now less pronounced. Accordingly, we shall only note the general formulation here.

For (2.14) as the primal problem, i.e., for

$$\min \{\Phi(x) = a_0^T x + x^T C x \mid Ax \le a_{.0}\}$$

we can formulate the dual problem as

$$\max \left\{\Phi(x, u) = a_0^T x + x^T C x + u^T(Ax - a_{.0}) \middle| \frac{\partial \Phi}{\partial x} = 0, u \ge 0\right\}.$$

If the relation

$$\frac{\partial \Phi}{\partial x} = a_{0.} + 2Cx + A^T u = 0$$

following from the constraints is substituted into the objective function, the dual program has the form

$$\max \{-x^T C x - a_{.0}^T u \mid 2Cx + A^T u = -a_{0.}, u \ge 0\}.$$

Note that for $C = 0$ the known duality formulation of the linear case is again obtained.

If one of these problems has a solution, the other does too, and the two extremal values are equal. It can also be shown (cf. Dorn [44]) that when $\hat{x}$ is a solution of the primal problem, it is also x-part of a solution of the dual problem. The converse is not generally true except when C is strictly definite. In that case, the x-part of a solution of the dual problem represents at the same time the solution of the primal problem.

As mentioned earlier, no general algorithms for the nonlinear optimization exist at this time. The best methods known so far are for the case of quadratic optimization and some of these will be explained in greater detail in the following sections.

2.7   The Method of Beale

The Beale method [6] belongs to the geometrically most illustrative methods for quadratic optimization. We begin with formulation (2.13) which requires the minimization of a convex quadratic objective function

$$Q(x_1, \ldots, x_n) = Q(x)$$

subject to the linear constraints

$$Ax = a_{.0}$$
$$x \geq 0. \tag{2.19}$$

The Beale method is started with any feasible basic solution of system (2.19). We solve the system of equations in (2.19) with respect to the chosen basic variables; if these are assumed to be the first m variables, i.e., $x_1, \ldots, x_m$, this leads to

$$x_g = d_{g0}^1 + \sum_{h=1}^{n-m} d_{gh}^1 z_h \qquad (g = 1, \ldots, m) \tag{2.20}$$

where

$$z_h = x_{m+h}$$

(the upper index 1 refers to the first approximation). Because of our particular choice, the basic variables assume the value $d_{g0}^1 \geq 0$ at the initial point of approximation. As in the case of linear optimization, we call the variables on the right side of (2.20) the "independent" or "vanishing" variables. Those on the left side are then the "dependent" or basic variables.

Using (2.20) we can eliminate the basic variables from Q. For the sake of simplicity the following notation is recommended:

$$Q(x_1, \ldots, x_n) = Q^1(z_1, \ldots, z_{n-m})$$

$$= c_{00}^1 + 2\sum_{i=1}^{n-m} c_{i0}^1 z_i + \sum_{h=1}^{n-m}\sum_{i=1}^{n-m} c_{hi}^1 z_i z_h$$

$$= c_{00}^1 + \sum_{i=1}^{n-m} c_{0i}^1 z_i + \sum_{h=1}^{n-m}\left(c_{h0}^1 + \sum_{i=1}^{n-m} c_{hi}^1 z_i\right) z_h$$

$$= (c_{00}^1 + c_{01}^1 z_1 + \cdots + c_{0,\,n-m}^1 z_{n-m}) \cdot 1$$

$$\quad + (c_{10}^1 + c_{11}^1 z_1 + \cdots + c_{1,\,n-m}^1 z_{n-m}) z_1$$

$$\cdot$$
$$\cdot$$
$$\cdot$$

$$\quad + (c_{h0}^1 + c_{h1}^1 z_1 + \cdots + c_{h,\,n-m}^1 z_{n-m}) z_h$$

$$\cdot$$
$$\cdot$$
$$\cdot$$

$$\quad + (c_{n-m,0}^1 + c_{n-m,1}^1 z_1 + \cdots + c_{n-m,\,n-m}^1 z_{n-m}) z_{n-m}.$$

$$\tag{2.21}$$

In (2.21) the symmetry $c_{ih}^1 = c_{hi}^1$ holds, and we further have

$$\frac{1}{2}\frac{\partial Q^1}{\partial z_h} = c_{h0}^1 \qquad (h = 1, \dots, n - m).$$

(2.22)

Clearly, the value of Q at the first approximation is equal to c_{00}^1. With the above notation the Kuhn–Tucker conditions assume a particularly simple form. In fact, if we have at the chosen trial point

$$\frac{\partial Q^1}{\partial z_h} \geq 0,$$

then this point already represents the optimal solution, since every

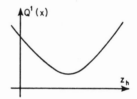

Fig. 11. For $z_h = 0$ we have $\partial Q^1/\partial z_h < 0$.

increase of the independent variables would increase the value of Q^1. However, if for certain z_h

$$\frac{\partial Q^1}{\partial z_h} < 0, \qquad \text{i.e.,} \qquad c_{h0}^1 < 0,$$

(2.23)

holds at the trial point, then, as Fig. 11 indicates, it is possible to improve the Q-value by making z_h positive.

Suppose this happens for $h = 1$, i.e., for z_1. Then, if z_1 increases, the other dependent variables will of course also change. As in the case of linear optimization, the question now arises how much the variable z_1 should increase. In the quadratic case we have to distinguish between the following two possibilities:

CASE 1. Let z_1 increase until one of the basic variables disappears as in the case of linear optimization (see Fig. 12).

CASE 2. $\partial Q^1/\partial z_1$ becomes zero before one of the dependent variables does in which case z_1 is of course increased only until $\partial Q^1/\partial z_1 = 0$ (Fig. 13), since otherwise the value of the objective function would begin to increase again.

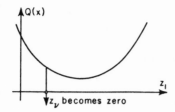

Fig. 12. z_v becomes zero before $\partial Q^1/\partial z_1$ does.

In case 1 the exchange step proceeds as in the case of linear optimization (step 4 in 1.2), and the constraint system has to be solved again for the new basic variables. These are then substituted into the objective function and with this the second approximation has been obtained.

In case 2 we introduce a new variable u_1 by

$$u_1 = \frac{1}{2} \frac{\partial Q^1}{\partial z_1}. \tag{2.24}$$

u_1 is not sign-restricted and is called the first "free" variable. As second approximation we choose that point at which the first free variable disappears together with all independent variables except that one which has entered the basis, i.e., except z_1. Here, too, the constraint system and the objective function are newly rearranged whereby the free variable u_1 is included among the independent variables. Because of

$$u_1 = c_{10}^1 + \sum_{h=1}^{n-m} c_{1h}^1 z_h = \frac{1}{2} \frac{\partial Q^1}{\partial z_1}, \tag{2.25}$$

we find for the now dependent variable z_1 that

$$z_1 = -\frac{c_{10}^1}{c_{11}^1} + \frac{1}{c_{11}^1} u_1 - \sum_{h=2}^{n-m} \frac{c_{1h}^1}{c_{11}^1} z_h = d_{10}^2 + d_{11}^2 u_1 + \sum_{h=2}^{n-m} d_{1h}^2 z_h.$$

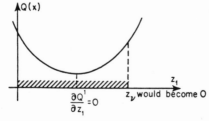

Fig. 13. $\partial Q^1/\partial z_1$ becomes zero before z_v does.

If this equation is used to eliminate z_1 from (2.20), it follows that

$$x_g = d_{g0}^2 + d_{g1}^2 u_1 + \sum_{h=2}^{n-m} d_{gh}^2 z_h \qquad (g = 1, 2, \ldots, m, m + 1). \quad (2.26)$$

In the same way, z_1 is also eliminated from Q^1.

The new constraint system (2.26) has one additional equation which arose from the introduction of the free variable. We also have one additional basic variable since no former basic variable has disappeared.

With this second approximation we proceed once again with the same rules as before. It should be noted that the Kuhn–Tucker condition for a free variable has the form

$$\frac{\partial Q^2}{\partial u_1} = 0.$$

If the derivative with respect to a free variable is not zero, then Q can be lowered by varying u_1 positively or negatively. More precisely, if the derivative with respect to u_1 is larger than zero, u_1 has to become negative and vice versa.

If a free variable has become dependent it need not be considered any longer once it has been eliminated from the constraint equations and from the objective function. (In fact, the only purpose of these equations is to assure that the variables do not become negative; such a control is not necessary for free variables.) Beale [6] has proved that this method leads in finitely many steps to the desired optimum provided the following additional rule is used.

Supplementary Rule. Whenever possible, the free variables shall be changed first when proceeding to the next approximation. Only if the derivatives of Q with respect to all the free variables are zero, and hence if in this way a decrease of Q is not possible, shall one of the original variables be introduced into the basis.

For the finiteness proof we refer to Beale [6] and Künzi and Krelle [95].

Example 10. With the help of the Beale method minimize

$$Q = -x_1 - 2x_2 + \tfrac{1}{2}x_1^2 + \tfrac{1}{2}x_2^2$$

subject to the constraints

$$2x_1 + 3x_2 + x_3 = 6, \quad x_1 + 4x_2 + x_4 = 5, \quad x_i \geq 0, \quad i = 1, \ldots, 4.$$

A feasible basic solution is given by $x_3 = 6$ and $x_4 = 5$, i.e., $x_1 = x_2 = 0$. With this we obtain for the system (2.20)

$$x_3 = 6 - 2x_1 - 3x_2, \qquad x_4 = 5 - x_1 - 4x_2.$$

Equation (2.21) then has the form

$$Q^1 = (0 - \tfrac{1}{2}x_1 - x_2) \cdot 1 + (-\tfrac{1}{2} + \tfrac{1}{2}x_1 + 0)x_1 + (-1 + 0 + \tfrac{1}{2}x_2)x_2.$$

Accordingly, we have at the initial approximation point

$$Q^1 = c_{00}^1 = 0, \qquad \frac{\partial Q_1}{\partial x_2} = 2c_{20} = -2.$$

The optimization has not been reached yet and x_2 has to become positive, i.e., x_2 has to be exchanged with x_4. At the same time $\partial Q^1/\partial x_2 < 0$ has to remain valid. We find

$$x_2 = \tfrac{5}{4} - \tfrac{1}{4}x_1 - \tfrac{1}{4}x_4, \qquad x_3 = \tfrac{9}{4} - \tfrac{5}{4}x_1 + \tfrac{3}{4}x_4$$

and therefore

$$Q^2 = (-\tfrac{55}{32} - \tfrac{13}{32}x_1 + \tfrac{3}{32}x_4) \cdot 1$$
$$+ (-\tfrac{13}{32} + \tfrac{17}{32}x_1 + \tfrac{1}{32}x_4)x_1$$
$$+ (\ \ \tfrac{3}{32} + \tfrac{1}{32}x_1 + \tfrac{1}{32}x_4)x_4.$$

Now $\partial Q^2/\partial x_1 = -\tfrac{13}{16} < 0$, i.e., x_1 has to become positive. We have

$$x_2 = 0 \quad \text{for} \quad x_1 = 5 \qquad \text{and} \qquad x_3 = 0 \quad \text{for} \quad x_1 = \tfrac{9}{5}$$

and

$$\frac{1}{2}\frac{\partial Q^2}{\partial x_1} = -\tfrac{13}{32} + \tfrac{17}{32}x_1 + \tfrac{1}{32}x_4 = 0 \qquad \text{for} \quad x_1 = \tfrac{13}{17}.$$

x_1 is not permitted to increase as far as the constraints would allow but only up to the value $x_1 = \tfrac{13}{17}$. Accordingly, we introduce the artificial variable $u_1 = (\tfrac{1}{2})\,\partial Q^2/\partial x_1$ and select as third approximation $u_1 = 0$, $x_4 = 0$. In other words, x_1 is newly inserted into the basis, and we find

$$x_1 = \tfrac{13}{17} + \tfrac{32}{17}u_1 - \tfrac{1}{17}x_4$$
$$x_2 = \tfrac{18}{17} - \tfrac{8}{17}u_1 - \tfrac{4}{17}x_4$$
$$x_3 = \tfrac{22}{17} - \tfrac{40}{17}u_1 + \tfrac{16}{17}x_4$$

and therefore

$$Q^3 = \ \ (-\tfrac{69}{34} \qquad\qquad + \tfrac{2}{17}x_4) \cdot 1$$
$$+ (\ \ \ 0 + \tfrac{32}{17}u_1 \qquad\quad)u_1$$
$$+ (\ \ \ \tfrac{2}{17} \qquad\quad + \tfrac{1}{34}x_4)x_4.$$

Now $\partial Q^3/\partial u_1 = 0$ and $\partial Q^3/\partial x_4 \geq 0$, and hence the optimum has been reached and the solution is:

$$x_1 = \tfrac{13}{17}, \qquad x_2 = \tfrac{18}{17}, \qquad x_3 = \tfrac{22}{17}, \qquad x_4 = 0, \qquad Q = -\tfrac{69}{34}.$$

2.8   The Method of Wolfe[1]

The method of Wolfe is distinguished by the fact that it works extensively with the simplex method and that it became accessible to electronic computers at a very early stage. The method begins with system (2.18), i.e., a solution is sought for

$$\left. \begin{array}{c} Ax = a_{.0} \\[6pt] 2Cx - v + A^T u = -a_{0.} \\[6pt] x \geq 0, \quad v \geq 0 \end{array} \right\} \tag{2.27}$$

$$x^T v = 0 \tag{2.28}$$

The values a_{j0} in (2.27) are assumed to be nonnegative. It should be noted once again that by definition x and v are n-vectors, while u is an m-vector. System (2.27) represents $m + n$ equations in $2n$ sign-restricted and m unrestricted variables.

The constraint $x^T v = 0$ implies that for no index i can v_i and x_i be positive simultaneously. Therefore, among the $2n$ variables x and v there can be at most n positive ones; in other words, of the $2n + m$ variables at least n have to disappear if the condition (2.28) is to be satisfied.

As in the case of linear optimization, a solution with at least n zero variables is called a basic solution. It is sufficient to find one among the basic solutions of the system for which the condition (2.28) is satisfied. An obvious move for bringing this about would be to return to the simplex method of linear optimization in some slightly modified form, since this method is concerned with the exchange of bases for a linear system of equations.

In the Wolfe method additional variables are introduced into system (2.27) in such a way that a feasible basis solution can be given immediately for which condition (2.28) is satisfied. It is then up to the simplex method to let these additional variables disappear again. Care must be taken during the iterative process that the additional condition (2.28) remains satisfied.

The algorithm can be given in two forms, a short and a long one. The long form works without restrictions; for the short one either $a_{0.} = 0$ or C has to be positive definite.

[1] Cf. Wolfe [158]

2.8.1 THE SHORT FORM

In order to find a solution for (2.27) we introduce $m + 2n$ additional nonnegative slack variables, namely:

$$w^T = (w_1, \ldots, w_m)$$
$$z^1 = (z_1^1, \ldots, z_n^1)$$
$$z^2 = (z_1^2, \ldots, z_n^2).$$

Then (2.27) is extended to a system of $m + n$ equations in $4n + 2m$ variables

$$Ax \qquad\qquad\qquad + w = a_{.0}$$
$$2Cx - v + A^T u + z^1 - z^2 = -a_{0.} \qquad (2.29)$$

with

$$x \geq 0, \quad v \geq 0, \quad w \geq 0, \quad z^1 \geq 0, \quad z^2 \geq 0.$$

Now a basic solution with $(4n + 2m) - (m + n) = 3n + m$ disappearing variables can be given immediately, which satisfies condition (2.28), namely,

$$x = 0, \qquad v = 0, \qquad u = 0$$

where for each index i at least one of the two variables

$$z_i^1 \qquad \text{or} \qquad z_i^2$$

is zero. Then the first basis includes the variables

$$w_j = a_{j0} \qquad \text{(nonnegative by assumption)}$$

for $j = 1, \ldots, m$ and for each i one of the variables z_i^1 or z_i^2, or more specifically,

$$z_i^1 = -a_{0i} \qquad \text{in case} \quad a_{0i} \quad \text{is negative}$$
$$z_i^2 = +a_{0i} \qquad \text{in case} \quad a_{0i} \quad \text{is positive}.$$

The reduction of the slack variables to the value 0 which leads from system (2.29) back to (2.27), now proceeds in two phases. In the first phase, starting from the first basic solution, the ordinary simplex method of linear optimization is used in order to minimize the linear form

$$\sum_{j=1}^{m} w_j$$

subject to the constraints of (2.29) as well as to the additional conditions

$$u = 0 \qquad \text{and} \qquad v = 0.$$

Accordingly, during the first phase the variables u_j and v_j remain outside the basis. If the constraints of system (2.27) are not incompatible, Wolfe [158] has proved that the minimum for $\sum_{j=1}^{m} w_j$ reaches the value 0. At the conclusion of the first phase we have found a basic solution of the system:

$$Ax \qquad\qquad\qquad = a_{.0}$$
$$2Cx - v + A^T u + Dz = -a_{0.} \qquad\qquad (2.30)$$
$$x \geq 0, \qquad v \geq 0, \qquad z \geq 0.$$

In (2.30) D is a diagonal matrix with either $+1$ or -1 on the diagonal, depending on whether $z_i^{\,1}$ or $z_i^{\,2}$ remain.

With this basic solution of (2.30) the second phase begins and the simplex method is used to minimize the linear form $\sum_{i=1}^{n} z_i$ subject to the constraints (2.30).

It can be proved (cf. Künzi and Krelle [95]) that the linear form $\sum_{i=1}^{n} z_i$ can be reduced to zero while condition (2.28) remains satisfied provided that, as mentioned above, either $a_{0.} = 0$ or C is positive definite.

With this a solution for the system (2.27)-(2.28) has been found and along with it a solution of the originally formulated problem (2.13).

2.8.2   THE LONG FORM

The long form consists of three phases, the first two of which are essentially the same as those of the short form. In principle, one proceeds by first applying the short form, but in such a way that the vector $a_{0.}$ is replaced by the zero vector. This means that in (2.29) the equation

$$2Cx - v + A^T u + z^1 - z^2 = -a_{0.}$$

is replaced by

$$2Cx - v + A^T u + z^1 - z^2 = 0.$$

As noted above, the short form then leads to the solution in two phases, and at their conclusion we are in possession of a basic solution of the system

$$Ax \qquad\qquad\qquad = a_{.0}$$
$$2Cx - v + A^T u = 0 \qquad\qquad (2.31)$$
$$x \geq 0, \qquad v \geq 0$$

which satisfies condition (2.28).

Now an additional variable μ is introduced by setting

$$Ax \qquad\qquad = a_{.0}$$
$$2Cx - v + A^T u + \mu a_{0.} = 0 \qquad\qquad (2.32)$$
$$x \geq 0, \qquad v \geq 0, \qquad \mu \geq 0.$$

The last basic solution is still valid for $\mu = 0$. What we need now is a basic solution for $\mu = 1$ satisfying (2.28); this will then be the optimal solution of problem (2.27) and therefore of (2.13).

To obtain this solution we proceed to the already mentioned third phase. Starting from the basic solution established in phase two, we use the third phase to minimize the objective function, now consisting only of a linear term, namely $-\mu$, under the constraints (2.32) and (2.28). During this optimization two cases can arise:

 i. $-\mu$ decreases indefinitely;
 ii. a bounded optimum exists for $-\mu$.

It can be shown that when $-\mu$ cannot be reduced to $-\infty$ during the third phase, it is not possible to execute even one step of the simplex method without violating the condition (2.28). (Cf. Wolfe [158] or Künzi and Krelle [95].)

Let (x^i, v^i, u^i, μ^i) be a sequence of iterates obtained in phase three; we are looking for a solution with $\mu = 1$. It is found as a linear combination of two solutions. Suppose the sequence of the μ-values is:

$$0 = \mu^1 < \mu^2 < \mu^3 < \cdots < \mu^g.$$

Two cases are then possible:

1. $\mu^g \geq 1$. Choose an index j with $1 \leq j \leq g - 1$ such that $\mu^j < 1 \leq \mu^{j+1}$. Then

$$(\hat{x}, \hat{v}, \hat{u}, \hat{\mu}) = \frac{\mu^{j+1} - 1}{\mu^{j+1} - \mu^j}(x^j, v^j, u^j, \mu^j) + \frac{1 - \mu^j}{\mu^{j+1} - \mu^j}(x^{j+1}, v^{j+1}, u^{j+1}, \mu^{j+1}).$$
$$(2.33)$$

Since $\mu = 1$, (2.33) represents a solution of problem (2.13).

2. $\mu^g < 1$. Form

$$(\hat{x}, \hat{v}, \hat{u}, \hat{\mu}) = (x^g, v^g, u^g, \mu^g) + \frac{1 - \mu^g}{\mu^{g+1}}(x^{g+1}, v^{g+1}, u^{g+1}, \mu^{g+1}). \quad (2.34)$$

Here again $\mu = 1$, which implies that the x-part of problem (2.13) has been solved.

This describes the long form in full.

Example 11. Solve Example 10 with the help of the Wolfe method. System (2.27)–(2.28) then has the form

$$2x_1 + 3x_2 + x_3 \qquad\qquad = 6$$
$$x_1 + 4x_2 \qquad\quad + x_4 = 5$$
$$x_1 - v_1 + 2u_1 + u_2 = 1$$
$$x_2 - v_2 + 3u_1 + 4u_2 = 2$$
$$- v_3 + u_1 \qquad\qquad = 0$$
$$- v_4 \qquad + u_2 = 0$$
$$x_1, x_2, x_3, x_4 \geq 0 \qquad v_1, v_2, v_3, v_4 \geq 0$$
$$x_1 v_1 + x_2 v_2 + x_3 v_3 + x_4 v_4 = 0.$$

The long form of the algorithm is used for the solution.

After introducing the nonnegative slack variables w, z^1, z^2 and μ we obtain the extended system (2.29):

$$2x_1 + 3x_2 + x_3 \qquad\qquad\qquad + w_1 \qquad = 6$$
$$x_1 + 4x_2 \qquad + x_4 \qquad\qquad + w_2 \qquad = 5$$
$$x_1 - v_1 \qquad + 2u_1 + u_2 + z_1^1 - z_1^2 \qquad - \mu = 0$$
$$x_2 - v_2 \qquad + 3u_1 + 4u_2 + z_2^1 - z_2^2 \qquad - 2\mu = 0$$
$$- v_3 \qquad + u_1 \qquad + z_3^1 - z_3^2 \qquad = 0$$
$$- v_4 \qquad + u_2 + z_4^1 - z_4^2 \qquad = 0$$
$$x_1 v_1 + x_2 v_2 + x_3 v_3 + x_4 v_4 = 0.$$

During the first phase $w_1 + w_2$ are minimized whereby μ, u, and v do not enter the basis and hence remain zero (in our calculation $-\sum w_i$ was maximized). A first feasible basis of this system consists of w and z^1, and accordingly the initial tableau reads:

		$-x_1$	$-x_2$	$-x_3$	$-x_4$	$-v_1$	$-v_2$	$-v_3$	$-v_4$	$-u_1$	$-u_2$	$-z_1^2$	$-z_2^2$	$-z_3^2$	$-z_4^2$	$-\mu$
w_1	6	2	3	1												
w_2	5	1	4		1											
z_1^1	0	1				-1				2	1	-1				-1
z_2^1	0		1				-1			3	4		-1			-2
z_3^1	0							-1		1				-1		
z_4^1	0								-1		1				-1	
$-\sum w_i$	-11	-3	-7	-1	-1											

After four exchange steps the final tableau of phase one is given by

		$-z_1^1$	$-z_2^1$	$-w_1$	$-x_4$	$-v_1$	$-v_2$	$-v_3$	$-v_4$	$-u_1$	$-u_2$	$-z_1^2$	$-w_2$	$-z_3^2$	$-z_4^2$	$-\mu$
x_3	$\tfrac{9}{4}$	$-\tfrac{5}{4}$		1	$-\tfrac{3}{4}$	$\tfrac{5}{4}$				$-\tfrac{5}{2}$	$-\tfrac{5}{4}$	$\tfrac{5}{4}$	$-\tfrac{3}{4}$			$\tfrac{5}{4}$
z_2^2	$\tfrac{5}{4}$	$-\tfrac{1}{4}$	-1		$\tfrac{1}{4}$	$\tfrac{1}{4}$	1			$-\tfrac{7}{2}$	$-\tfrac{17}{4}$	$\tfrac{1}{4}$	$\tfrac{1}{4}$			$\tfrac{9}{4}$
x_1	0	1				-1				2	1	-1				-1
x_2	$\tfrac{5}{4}$	$-\tfrac{1}{4}$			$\tfrac{1}{4}$	$\tfrac{1}{4}$				$-\tfrac{1}{2}$	$-\tfrac{1}{4}$	$\tfrac{1}{4}$	$\tfrac{1}{4}$			$\tfrac{1}{4}$
z_3^1	0							-1		1				-1		
z_4^1	0								-1	$-\tfrac{5}{2}$	1				1	
$-\Sigma w_i$	0			1									1			1

In phase two $\sum z_i$ is minimized or $-\sum z_i$ maximized subject to the additional condition that x_i and v_i are not allowed to be in the basis simultaneously. Furthermore, μ is not permitted to enter the basis either. In order to obtain the starting tableau of phase two, we delete the columns belonging to w_i and to those z_i^1 or z_i^2 not contained in the basis. We then obtain:

		$-x_4$	$-v_1$	$-v_2$	$-v_3$	$-v_4$	$-u_1$	$-u_2$	$-\mu$
x_3	$\tfrac{9}{4}$	$-\tfrac{3}{4}$	$\tfrac{5}{4}$				$-\tfrac{5}{2}$	$-\tfrac{5}{4}$	$\tfrac{5}{4}$
z_2	$\tfrac{5}{4}$	$\tfrac{1}{4}$	$\tfrac{1}{4}$	1			$-\tfrac{7}{2}$	$-\tfrac{17}{4}$	$\tfrac{9}{4}$
x_1	0		-1				2	1	-1
x_2	$\tfrac{5}{4}$	$\tfrac{1}{4}$	$\tfrac{1}{4}$				$-\tfrac{1}{2}$	$-\tfrac{1}{4}$	$\tfrac{1}{4}$
x_3	0			-1			1		
z_4	0				-1	$-\tfrac{5}{2}$		1	
$-\Sigma z_i$	$\tfrac{5}{4}$	$-\tfrac{1}{4}$	$-\tfrac{1}{4}$	-1	1	1	$\tfrac{5}{2}$	$\tfrac{13}{4}$	$-\tfrac{9}{4}$

and from this, after three exchange steps, the final tableau of phase two has been reached:

		$-z_2$	$-v_1$	$-v_2$	$-v_3$	$-v_4$	$-z_3$	$-z_4$	$-\mu$
x_3	6	3	2	3	-13	-14	13	14	8
x_4	5	4	1	4	-14	-17	14	17	9
x_1	0		-1		2	1	-2	-1	-1
x_2	0	-1		-1	3	4	-3	-4	-2
u_1	0				-1		1		
u_2	0					-1		1	
$-\Sigma z_i$	0	1					1	1	

Now we are in the possession of a basic solution of system (2.31) which satisfies condition (2.28) and forms the starting point for phase three. Accordingly, the two z-columns are deleted and $+\mu$ is maximized under continuing consideration of the additional condition (2.28). The initial tableau then reads:

		$-v_1$	$-v_2$	$-v_3$	$-v_4$	$-\mu$
x_3	6	2	3	-13	-14	8
x_4	5	1	4	-14	-17	9
x_1	0	-1		2	1	-1
x_2	0			3	4	-2
u_1	0		-1	-1		
u_2	0				-1	
μ	0					-1

After the first exchange step we find

$$\mu^2 = \tfrac{5}{9} < 1 \quad \text{and} \quad x_1{}^2 = \tfrac{5}{9}\,; \quad x_2{}^2 = \tfrac{10}{9}\,; \quad x_3{}^2 = \tfrac{14}{9}\,; \quad x_4{}^2 = 0\,;$$

and after the second one

$$\mu^3 = \tfrac{16}{5} > 1 \quad \text{and} \quad x_1{}^3 = \tfrac{9}{5}\,; \quad x_2{}^3 = \tfrac{4}{2}\,; \quad x_3{}^2 = 0\,; \quad x_4{}^2 = 0$$

and with it the following tableau:

		$-v_1$	$-v_2$	$-v_3$	$-x_3$	$-x_4$
v_4	$\tfrac{7}{5}$	1	$-\tfrac{1}{2}$	$-\tfrac{1}{2}$	$\tfrac{9}{10}$	$-\tfrac{4}{5}$
μ	$\tfrac{16}{5}$	2	$-\tfrac{1}{2}$	$-\tfrac{5}{2}$	$\tfrac{17}{10}$	$-\tfrac{7}{5}$
x_1	$\tfrac{9}{5}$				$\tfrac{4}{5}$	$-\tfrac{3}{5}$
x_2	$\tfrac{4}{5}$				$-\tfrac{1}{5}$	$\tfrac{2}{5}$
u_1	0			-1		
u_2	$\tfrac{7}{5}$	1	$-\tfrac{1}{2}$	$-\tfrac{1}{2}$	$\tfrac{9}{10}$	$-\tfrac{4}{5}$
μ	$\tfrac{16}{5}$	2	$-\tfrac{1}{2}$	$-\tfrac{5}{2}$	$\tfrac{17}{10}$	$-\tfrac{7}{5}$

With $\mu^g = \mu^3 \geq 1$ and $\mu^j = \mu^2 < 1 \leq \mu^{j+1} = \mu^3$ we then obtain

$$(x, \hat{v}, \hat{u}, \hat{\mu}) = \tfrac{99}{119}(x^2, v^2, u^2, \mu^2) + \tfrac{20}{119}(x^3, v^3, u^3, \mu^3)$$

and therefore as final solution

$$\hat{x}_1 = \tfrac{13}{17}, \quad \hat{x}_2 = \tfrac{18}{17}, \quad \hat{x}_3 = \tfrac{22}{17}, \quad \hat{x}_4 = 0.$$

As a control we can compute that indeed $\hat{\mu} = 1$.

2.9 A Look at Further Methods

In conclusion of this chapter we shall briefly mention three other methods deserving attention from a mathematical viewpoint, all the more so since they appear to be extendable, particularly where nonlinear constraints are concerned.

THE GRADIENT METHOD OF FRISCH, ROSEN, AND ZOUTENDIJK[1]

Although the earlier methods are to some extent also gradient methods, the above-listed procedures are said to be gradient methods in the proper sense of the word since they determine the gradient direction at each step of the iteration in order to get the optimal increase.

A comparison of the different methods shows that all of them have certain procedural similarities. If, for example, a concave objective function is to be maximized on a domain R, represented by

$$a_j{}^T x \leq b_j,$$

it is assumed that the function $Q(x)$ has a continuous gradient in R. Restricting ourselves to the so-called methods with long steps, we start with an arbitrary point x^0, located on the boundary or in the interior of R. In order to proceed from the kth iteration point x^k to x^{k+1}, a direction s^k is determined at x^k in such a way that for small $\lambda > 0$ the ray $x^k + \lambda s^k$ continues to be in R. For this it is necessary and sufficient that

$$a_j{}^T s^k \leq 0 \qquad \text{for all } j \text{ with} \qquad a_j{}^T x^k = b_j. \qquad (2.35)$$

Such a direction is called admissible. Furthermore, the Q value has to increase along the ray at least for small λ; this provides the condition

$$s^{k^T} g(x^k) > 0$$

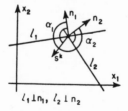

Fig. 14. Geometrical interpretation of the gradient methods in the plane.

[1] From Refs. 59, 123, and 164.

where $g(x^k)$ denotes the gradient of the objective function at x^k. Such an s^k is called usable (Fig. 14).

Conditions (2.35) state that α_1 and α_2 have to be not less than 90°. The different methods consist in determining, by means of suitable rules, an s^k from among the set of all admissible and usable directions s at x^k which is as favorable as possible. Once such an s^k has been found, the computation of the finite step length λ^k is the same for all methods:

The value λ' is determined at which the ray leaves the domain, as is the value λ'' at which Q is optimal on the ray. Then λ^k is given by

$$\lambda^k = \min (\lambda', \lambda''),$$

and we have

$$x^{k+1} = x^k + \lambda^k s^k.$$

Frisch [59] uses the so-called multiplex method in which two different directions are utilized. Either the iteration moves in a direction vertical to a certain boundary (in order to enter the interior) or it moves inside such a boundary. Closely related to the Frisch method is the projected gradient method of Rosen [123]. In it, the gradient at the point x^k is projected onto the intersection of the set of all hyperplanes on which x^k is located. Then the iteration proceeds along the projected direction, provided the dimension of the intersection of the hyperplane was not zero. A decision has to be made at each point x^k whether, in the interest of improving the objective function, the intersection of the hyperplanes has to be changed or not.

Recently, Zoutendijk [164] has added to the already known methods a very effective new one called the method of feasible directions. The process begins with a feasible initial solution and attempts to find a solution for an improved objective function. For this purpose Zoutendijk determines the so-called best direction. This reduces in each case to the solution of a small linear or nonlinear auxiliary program which can always be solved with the help of the simplex method. For the determination of the best direction the linear form

$$g^T s$$

is maximized subject to the constraints

$$a_j^T s \leq 0,$$

where $g(x^k) = g$ represents the gradient at the point x^k. In the case of a quadratic objective function, the method can be made finite by using the theory of conjugate gradients of Hestenes and Stiefel [75].

All three mentioned gradient methods work for semi-definite forms and are therefore rather general. Furthermore, it is also possible to use these methods in the case of nonlinear constraints. For additional methods in the theory of nonlinear optimization we refer to Künzi and Krelle [95].

3 EXPLANATIONS OF
THE COMPUTER PROGRAMS

3.1 The Subroutine System

The first two, theoretically oriented, chapters of this book provide self-contained presentations of different methods for linear and quadratic optimization. Chapter 4 (the program part) contains the corresponding computer programs in ALGOL and FORTRAN IV. While didactic considerations determined the emphasis in the theoretical discussion, stress was placed in the programs on effective machine utilization with respect to storage requirements and computation time. As a result, certain differences between the two parts of the book are inevitable in the presentation of the methods. This chapter will explain those differences, while at the same time providing detailed instructions for the use of the programs.

In all methods of mathematical optimization, certain fundamental computational steps occur again and again. These have therefore been programmed as global procedures and are accessible to all the individual optimization programs. The following global procedures are given in 4.1 where comments about particular details can also be found.

mp1 Ordering of a list
mp2 Determination of the minimum of a list of quotients
mp3 Exchange of a basis- and a nonbasis-variable in a tableau
mp4 Treatment of degeneracies (only used in *mp2*)
mp5 Determination of the minimum of a row or column in a tableau
mp7 Determination of the maximum of a row or column in a tableau
mp8 Determination of the absolute maximum of a row or column in a tableau

mp9 Transformation of a quadratic form (for the Beale method only)

mp10 Stepwise computation of an inverse matrix (for the revised simplex method only)

These global procedures are assumed to be available (that is, declared) in the actual optimization programs, and they are called without further specification. The parameter list contained in a procedure call gives complete information about the present meaning of the individual variables.

The actual optimization programs themselves also have the form of procedures or subroutines; this permits us to disregard the theoretically uninteresting but frequently extensive and mostly machine-dependable input-output portions of the programs.

3.2 The Use of the Optimization Programs

The optimization procedures contained in the program section represent complete programs which will work for even the most general cases, and the user therefore need not check these programs in detail himself. Instead, he can utilize them as a compact working tool for his own problems. On the other hand, he is still faced with one task—to imbed the desired optimization procedure into a suitable framework in order to realize the three following objectives:

i. A storage organization appropriate to the needs of the particular program and the chosen method.

ii. The input of all data used in the method (e.g., from cards or tape).

iii. The output of all results in a form best suited to the user.

Such a driver program will, therefore, always have to have the structure: given at the top of the next page.

A smooth interaction of the input-output portion of the program with the optimization procedures of this book is of course possible only if the rules for the storage of the data here summarized are strictly maintained.

If the user of an optimization procedure does wish to obtain not only the final results for his problem but certain intermediate results as well, he will naturally have to enter into the optimization procedures themselves. In that case (probably rare), appropriate output statements in ALGOL or FORTRAN have to be inserted at suitable places inside the optimization procedures.

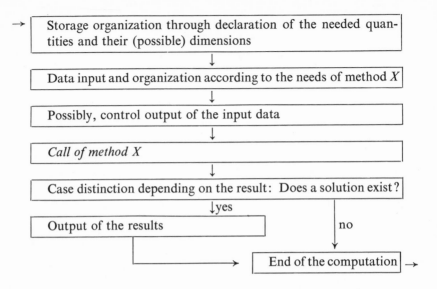

3.3   Numerical Properties

We have just spoken of the optimization procedures as generally valid programs which will handle all special cases and degeneracies automatically correctly. While this is theoretically correct, the practical aspects of numerical computations are unfortunately more complicated. Round-off errors and the representation of zero (depending on the type of computer) require the following considerations:

In most programs "checks for zero" occur; for example, in the determination of a feasible basis for the simplex method, a check has to be made whether the value of the auxiliary objective function is zero. Because of round-off errors this is generally not exactly satisfied. It is therefore desirable to introduce a number $\varepsilon > 0$ and to replace exact checks for zero by comparisons of ε with the modulus of the control variable, whereby ε would be an input variable.

We will not go into any theoretical considerations here concerning the computation of appropriate ε. In practice the following formula has proved to be very useful;

$$\varepsilon = \frac{\text{Sum of the moduli of the coefficients of } A}{\text{Number of the coefficients of } A} \cdot 10^{(-L+1)}$$

where L is the number of decimal digits in the mantissa of the floating point representation, and A the optimization matrix.

For larger problems it is recommended that double precision be used in those parts of the algorithms which are susceptible to round-off errors, as, for example, in the exchange method *mp3*.

However, the numerical stability of the different methods is quite different; particularly the revised form of the simplex method is in this respect significantly superior to the classical form; in fact, by appropriate follow-up inversions (see remark in 3.5.3) accumulated errors can even be eliminated in this method, provided this is at all possible, given the nature of the problem. Accordingly, appropriate selection of the method is frequently a refined way of assuring results with high accuracy.

The two techniques just mentioned (tolerance ε, double precision) are not included in the programs of this book. They would have caused the programs to differ even further from the presentations in the theoretical part; moreover, a skilled programmer can implement these measures without too much effort.

3.4 General Discussion of Variable Notations and Storage Organization

As much as possible, notations for the most important variables have been kept the same for all methods. A few brief descriptions follow below; those details which are particular to certain methods will be discussed separately in 3.5.

n	Number of independent variables (nonbasic variables)
m	Number of dependent variables (basic variables)
$a[\,.\,]$	"Simplex tableau" (corresponding to the matrix A in the theoretical section).

In the case of all linear programs (except for the decomposition algorithm), a always contains the coefficients of the constraints as well as of the objective function. The coefficients of the objective function thereby always stand in the 0th row, while the constant terms of the constraints form the 0th column of the matrix.

One computationally special approach is the selected storage organization of the actually *two*-dimensional tableau as a *simply subscripted* array whereby a matrix element is represented as follows:

Theory: a_{ik}[1]   Program: ALGOL: $a[i \times zschrt + k \times sschrt]$
FORTRAN: $A(i * ZSCHRT + K * SSCHRT + 1)$

This permits, without restoring, a very simple transposition of the tableau and of all operations to be applied to it. Moreover, the advantages of a columnwise or rowwise storage arrangement of a two-dimensional array are combined and the index calculation is nevertheless clear. The process is controlled by two auxiliary quantities:

zschrt[2]   Row step = distance in memory of two subsequent elements of a column

sschrt[3]   Column step = distance in memory of two subsequent elements of a row

For the values of zschrt and sschrt the following two combinations are possible:

Storage arrangement	zschrt	sschrt
"Columnwise"	1	$m + 1$
"Rowwise"	$n + 1$	1

The storage allocation for tableau A is, therefore:

in ALGOL: **array** $a[0 : (m + 1) \times (n + 1) - 1]$
in FORTRAN: DIMENSION $A(D)$ with $D = (M + 1) * (N + 1)$.

Because a vector cannot have a 0th element in FORTRAN, all indices in that case have to be raised by 1 (hence, $A(I * ZSCHRT + K * SSCHRT + 1)$). Exceptions to these rules—in the case of larger tableaus or of prescribed rowwise storage—are given in 3.5.

proto1[k], $k = 1, \ldots, n$ contains the indices of the nonbasic variables.
proto2[i], $i = 1, \ldots, m$ contains the indices of the basic variables.

For reasons of space, the identity matrix used in the theoretical part is not stored and is, therefore, not part of the coefficient array. This means that designation of the variables as to whether or not they belong to the basis has to be accomplished differently. For this purpose the original nonbasic variables are automatically given the indices 1 through n and the original basic variables the indices $n + 1$ through $n + m$. At the end of the calculation they are interchanged. For each position proto1 and

---

[1] In the theoretical part often designated by a_{ji}.
[2] Abbreviation for zeilenschritt.
[3] Abbreviation for spaltenschritt.

proto2 contain the original index of the variable now occupying it; hence, the results of the computation are obtained as follows:

variable in *proto1:*   present nonbasic variable; value is 0.

variable in *proto2:*   present basic variable; the value of the variable whose index is in *proto2*[i] is equal to the present a_{i_0}.

fall: this parameter of the result provides information at the end of the computation concerning the special case which arose in the treatment of the problem.

fall=0:   the problem *has a finite* solution

fall=1:   the problem has *no finite* solution

fall=2:   *no feasible* basic solution for the problem exists.

The nonuniqueness of the solution has not been indicated but can be recognized by the zeroes in the 0th row of the final tableau.

The following variables are not part of the parameter lists of the optimization methods but occur internally. Their description is therefore helpful only in understanding the programs.

l1[k]	($k = 1, \ldots, l1$[0]) contains the indices of the columns which are admissible for an exchange. *l1*[0] counts the number of these columns.
l2[i]	($i = 1, \ldots, l2$[0]) has the corresponding meaning for the row indices.
ip	the index of the pivot row
kp	the index of the pivot column
znr	a row index
q1	the minimal quotient which determines the pivot row
v	the sign of *q1*
i0, i1, k0, k1	denote at which row or column the exchange starts or ends
p1	indicates whether a transformation of the pivot row is necessary (1: yes; 0: no)
p2	indicates whether a transformation of the pivot column is necessary (1: yes; 0: no)

3.5 Properties of Individual Optimization Programs

3.5.1 THE SIMPLEX METHOD

(Compare: Theory: 1.2 and 1.3; Programs: ALGOL: 4.2, FORTRAN: 4.3.)

Linear inequalities and equations can be used in the simplex method and can have one of the three following forms:

$$a_{i0} + \sum_{k=1}^{n} a_{ik}x_k \geq 0 \qquad \text{with} \quad a_{i0} \geq 0 \qquad (3.1)$$

$$a_{i0} + \sum_{k=1}^{n} a_{ik}x_k \leq 0 \qquad \text{with} \quad a_{i0} \geq 0 \qquad (3.2)$$

$$a_{i0} + \sum_{k=1}^{n} a_{ik}x_k = 0 \qquad \text{with} \quad a_{i0} \geq 0 \qquad (3.3)$$

The particular problem is assumed to have $m1$, $m2$, or $m3$ constraints of the forms (3.1), (3.2), or (3.3), respectively. Hence, the total number of constraints is $m = m1 + m2 + m3$; for the input the following rowwise ordering has to be used in tableau A:

(1) Row 0:   Objective function (maximization problem)
(2) Rows 1 through $m1$:   Constraints of the form (3.1)
(3) Rows $m1 + 1$ through $m1 + m2$: Constraints of the form (3.2)
(4) Rows $m1 + m2 + 1$ through m: Constraints of the form (3.3)

Example 12. Maximize $(2x_1 + x_2)$ subject to the constraints $x_1 + x_2 \geq 2$, $x_1 + x_2 \leq 3$, and $x_1 = 1$. The array of coefficients then has the form:

column	0	1	2	
row				
0	0	2	1	$m = 3, n = 2$
1	3	−1	−1	$m1 = 1$
2	2	−1	−1	$m2 = 1$
3	1	−1	0	$m3 = 1$

STORAGE ALLOCATION. For internal reasons the driver program for the simplex method requires a storage area which is larger by one row than that required for the given constraints and the objective function. Hence, in the driver program the array must be declared as follows:

ALGOL: (the program requires an additional (-1)st row for A):

array $a[-n - 1 : (n + 1) \times (m + 1) - 1]$.

FORTRAN: (the program requires an additional $(m + 1)$st row in A):

DIMENSION A(D)   with $D = (N + 1) * (M + 2)$

COMPUTATIONAL APPROACH. In the case of constraints of the form (3.2) and (3.3) the program for the simplex method must first find a feasible basic solution. For this purpose the M-method is used and the auxiliary objective function is stored in that additional row of tableau A which was reserved for this purpose at storage allocation time. In the special case when only constraints of the form (3.1) are given, this part of the program is not used.

3.5.2 THE DUAL SIMPLEX METHOD

(Compare: Theory: 1.7; Programs: ALGOL: 4.4, FORTRAN: 4.5.)

Compared to the standard simplex method, the dual method is advantageous only under very special conditions, and we will therefore assume that these conditions are satisfied, namely:

i.   All constraints are inequalities (there are no equations).

ii. All coefficients of the objective function to be minimized are non-negative. (Such a simplex tableau is also said to be *dually feasible*.)

In case ii all constraints can be written in the form

$$a_{i0} + \sum_{k=1}^{n} a_{ik}x_k \geq 0 \qquad \text{with unrestricted} \quad a_{i0}. \qquad (3.4)$$

The simplex tableau contains the objective function (minimization problem) with nonnegative coefficients as row zero; then follow the m inequalities.

Example 13. Minimize $(x_1 + 3x_2)$ subject to the constraints $x_1 + x_2 \geq 2$ and $x_1 + 2x_2 \leq 3$. Then the tableau has the form:

column:	0	1	2	
row:				
0	0	1	3	$m = 2, n = 2$
1	−2	1	1	
2	3	−1	−2	

3.5.3 THE REVISED SIMPLEX METHOD

(Compare: Theory: 1.8. Programs: ALGOL 4.6; FORTRAN: 4.7.)

This method accomplishes the actual optimization very effectively and is little susceptible to numerical difficulties (round-off errors). The preparatory work, however, particularly the elimination of equations, requires use of the same techniques as in the case of the standard simplex method. We therefore assume that a feasible basic solution has already been determined and that only inequalities of the following form occur as constraints

$$a_{i0} + \sum_{k=1}^{n} a_{ik}x_k \geq 0 \quad \text{with} \quad a_{i0} \geq 0. \tag{3.5}$$

The tableau contains as row zero the objective function to be maximized and following it the m inequalities.

Example 14. Maximize $(2x_1 + x_2)$ subject to the constraints $x_1 + x_2 \leq 3$, $x_1 - x_2 \geq -1$; then the tableau has the form:

column	0	1	2	
row				
0	0	2	1	$m = 2, n = 2$
1	3	−1	−1	
2	1	1	−1	

COMPUTATIONAL APPROACH. Internally, the revised simplex method uses an inverse matrix B; with its help the constraints are recomputed at any step from the original values so that they constitute linear forms of the present nonbasic variables. This inverse B therefore requires additional storage space; but only on the order of m^2 locations, in contrast to the constraint matrix with its more than $m \cdot n$ locations. (In the case $n < m$, suitable dualization techniques permit the use of an inverse of size n^2.) It is assumed in the present program that both matrices A and B are in core memory. However, in case of lack of sufficient storage space, the program can easily be modified in such a way that only the inverse matrix B resides in core memory, while the larger constraint matrix A is kept in secondary storage (magnetic tape, discs, or drum). The revised simplex method has been written in such a way that such use of auxiliary storage influences the computational time only slightly.

In this method, an eradication of accumulated round-off is possible by recomputation of the matrix B directly from the initial data. This measure has not been incorporated into the present programs, but it can be implemented quite easily (for example, with the help of a standard program for "Gauss-Jordan inversion," probably available everywhere). It is more difficult to find a theoretically satisfactory, and at the same time practical, criterion, when such a re-inversion is to take place automatically. The simplest approach is a re-inversion after a fixed number of steps, a number to be supplied by the program user.

3.5.4 THE DECOMPOSITION ALGORITHM

(Compare: Theory: 1.9; Programs: ALGOL: 4.8, FORTRAN: 4.9.)

The form of the constraints for the decomposition algorithm has been explained in detail in (1.9). Tableau (1.70) presented there has to be stored for the computation in the two one-dimensional arrays a and b as follows:

$$
\begin{array}{|c|c|c|c|c|}
\hline
0 & c_1^T & c_2^T & \cdots & c_n^T \\
\hline
b & A_1 & A_2 & \cdots & A_n \\
\hline
\end{array}
\tag{3.6}
$$

$$
\left.
\begin{array}{cc}
\boxed{\begin{array}{c|c} b_1 & B_1 \end{array}} & \\
 & \boxed{\begin{array}{c|c} b_2 & B_2 \end{array}} \\
 & \qquad \ddots \\
 & \qquad \boxed{\begin{array}{c|c} b_n & B_n \end{array}}
\end{array}
\right\}
\tag{3.7}
$$

Summary of the Variables and Their Meaning

a	The entire tableau (3.6) (including its border rows and columns) stored in memory
zschrt, sschrt	Row and column step for a with the customary meaning
m	The number of constraints in tableau A
n	The number of partial forms in (3.7)
$i1$	Total number of rows ⎫ in both (3.6) and (3.7)
$i2$	Total number of columns ⎭
b	The entire tableau (3.7) consisting of the n partial forms (each of which contains the constant column b_k as 0th column). Storage of the partial forms within b is controlled by the following index vectors for $k = 1, \ldots , n$:
list1[k]	The row number of the kth partial form
list2[k]	The column number of the kth partial form
list3[k]	The row step of the kth partial form ⎫ Combinations as
list4[k]	The column step of the kth partial form ⎭ given in 3.4

list5[k]   Location of the initial element of the kth partial form in b. Actually, it is unimportant in which sequence and where the partial forms of the one-dimensional array b have been stored. However, in the interest of storage economy, the partial forms should be stored consecutively

i3   The largest row number of a partial form $= \max_k (list1[k])$

i4   The largest column number of a partial form $= \max_k (list2[k])$

proto[k]   Result-indicator: The number of columns of the kth partial form contained in the solution.

x   The solution vector

USE OF AUXILIARY PROGRAMS AS GLOBAL PROCEDURES

up   This subroutine designates a system for the solution of a linear optimization problem. For it we can substitute the program for the simplex method (cf. 4.2 and 4.3), the revised simplex method (cf. 4.6 and 4.7), or any corresponding method

pricevector   This subroutine recomputes the pricevector. Hence, it corresponds essentially to a step of the revised simplex method (cf. 4.6 and 4.7) and accordingly has not been reprogrammed here

COMPUTATIONAL APPROACH. The algorithm proceeds in strict agreement with the theoretical description in 1.9.

3.5.5 THE DUOPLEX METHOD

(Compare: Theory: 1.10; Programs: ALGOL: 4.10, FORTRAN: 4.11.)

For the duoplex method the constraints can be inequalities (3.8), as well as Eqs. (3.9):

$$a_{i0} + \sum_{k=1}^{n} a_{ik}x_k \geq 0 \qquad \text{with unrestricted} \quad a_{i0} \qquad (3.8)$$

$$a_{i0} + \sum_{k=1}^{n} a_{ik}x_k = 0 \qquad \text{with} \quad a_{i0} \geq 0. \qquad (3.9)$$

Consequently, the inequalities violated by the initial basic solution are recognizable by the occurrence of negative a_{i0}. The storage arrangement of the tableau A is assumed to be as follows:

(1) Row 0:   Objective function (maximization problem)
(2) Rows 1 through $m1$: $m1$ inequalities of the form (3.8)
(3) Rows $m1 + 1$ through $m1 + m2$: $m2$ equations of the form (3.9)

Example 15. Maximize $(2x_1 + x_2)$ subject to the constraints $x_1 + x_2 \leq 3$, $x_1 + x_2 \geq 2$, $x_1 = 1$; then the tableau has the form:

column row	0	1	2	
0	0	2	1	$m = 3, n = 2$
1	3	−1	−1	$m1 = 2$
2	−2	1	1	$m2 = 1$
3	1	−1	0	

COMPUTATIONAL APPROACH. At first the constraints of form (3.9) are removed from the basis. This diminishes the size of the tableau and in turn cuts down the computational work required for each step and in many cases also decreases the number of iteration steps. The feasible basic solution is found by means of the multiphase method whereby care is taken not to re-violate a constraint once it has been satisfied. (An auxiliary objective function is not required.)

For the sake of clarity, the program for the duoplex algorithm has been written in a form similar to that of the simplex method. This means that at each step of the iteration the entire tableau is transformed. As in the case of the revised simplex method, a "revised duoplex method" would show certain advantages over the original method, in particular with respect to numerical stability (in this connection, see the comments in 3.5.3).

3.5.6 THE GOMORY ALGORITHM

(Compare: Theory: 1.11; Programs: ALGOL: 4.12, FORTRAN: 4.13.)

The program for the Gomory algorithm can handle a tableau of constraints comprising inequalities as well as equations, but all independent variables are required to be integer valued.

MEANING OF THE VARIABLES AND EXAMPLE FOR THE TABLEAU. All parameters correspond to those of the simplex method (cf. 3.5.1).

STORAGE ALLOCATION. The tableau a must be stored rowwise ($zschrt = n + 1$; $sschrt = 1$).

During the course of the computation with the Gomory algorithm, the auxiliary constraints for maintaining integer values cause the tableau to increase. Therefore, the user's storage allocation declarations have to provide for a correspondingly larger tableau dimension (with maximally n additional rows).

USE OF SUBROUTINES AS GLOBAL PROCEDURES

simplex   the two procedures *simplex* and *dusex*
dusex   (dual simplex method described earlier) are assumed to be available. (For ALGOL, see program section 4.2 and 4.4; for FORTRAN, see 4.3 and 4.5.)

COMPUTATIONAL APPROACH. The Gomory algorithm is adjusted entirely toward the availability of the simplex method. Accordingly, it uses this method internally a number of times: In the first part the optimization problem is solved by means of *simplex* without considering the integer-value-condition for the original independent variables. If there is still a fractionally valued variable in this provisional solution which has to be given an integer value, then, in a second part, tableau a is expanded by an auxiliary constraint. This larger problem is treated once again with the (in this case dual) simplex method *dusex*. The second part is repeated as often as fractionally valued variables continue to exist among the original independent variables.

It is possible to estimate the number of auxiliary constraints which will be added during the process (see Gomory [64]), but this estimate, needed in case of very tight storage allocation, has but little practical significance as it is applicable (for other reasons) only to relatively small tableaus. Due to the nature of the auxiliary constraints, the method is very susceptible to round-off errors and for this reason practical experiences with the present program have been limited to examples containing few variables and few restrictions.

3.5.7 BEALE'S ALGORITHM

(Compare: Theory: 2.7; Programs: ALGOL: 4.14, FORTRAN: 4.15.)

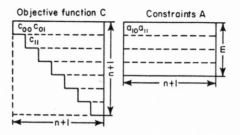

Fig. 15. Schematic representation of the storage of the objective function and the constraints in Beale's method.

This method determines the minimum of a definite quadratic form subject to linear constraints. The determination of a feasible basic solution of the constraint system has nothing to do with quadratic optimization, and is therefore assumed here to have been accomplished beforehand. As a result, the linear constraints will all have the form

$$a_{i0} + \sum_{k=1}^{n} a_{ik}x_k \geq 0 \qquad \text{with} \qquad a_{i0} \geq 0. \tag{3.10}$$

They are stored rowwise in the first through the mth row of tableau a. The 0th row of tableau a is not used since the symmetric quadratic objective function C requires an $(n + 1) \times (n + 1)$ array as storage area (and not just one row).

However, for reasons of storage economy, only the upper triangle of C is stored in the array c, in the form of densely packed row segments. The element c_{ik} therefore corresponds to $c[(2 \times n-i + 1) \times i/2 + k]$. This necessitates preparation of the two tableaus given in Fig. 15.

STORAGE ALLOCATION. Internally, the program expands tableau A by $(n + 1)$ additional rows. Hence this space has to be reserved when declaring A (but not for the input routine). In other words, the dimension of the storage area of A has to be $(m + n + 2) \times (n + 1)$.

COMPUTATIONAL APPROACH. Corresponding to each exchange of variables (dependent–independent) in the constraint tableau A, we have a corresponding exchange in the quadratic objective function C. This latter exchange is handled by means of the subroutine $mp9$. Because of C's triangular representation, program $mp9$ deviates somewhat from the theoretical discussion, but the comments included in $mp9$ (cf. 4.1) explain the approach in detail and always refer to Fig. 16. The other parts of the algorithm comply strictly with the theoretical discussion. Combination

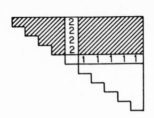

Fig. 16. Schematic representation of the quadratic form.

of the Beale method with the revised form of the simplex exchange offers advantages already mentioned earlier.

3.5.8   THE WOLFE ALGORITHM

(Compare: Theory: 2.8; Programs: ALGOL 4.16, FORTRAN: 4.17.)

This method computes the minimum of a quadratic form subject to linear constraints given in the form of equations:

$$\min \{c^T x + x^T C x \mid a_0 + Ax = 0, \quad a_0 \geq 0, \quad x \geq 0\}$$

All data for the constraints, the objective function and for all additional quantities described in the theoretical section are stored in one large tableau of the form (3.11).

keep row 0 free						1	
a_0	A $m \times n$	0 $m \times n$	0 $m \times m$	0 $m \times n$		o	m
o	$-2C$ $n \times n$	E $n \times n$	A^T $n \times m$	E $n \times n$		$-c_0$	n
1	n	n	m	n		1	

$\leftarrow m + n + 1 \text{ rows} \rightarrow$ (3.11)

$\leftarrow 3 \times n + m + 2 \text{ columns} \rightarrow$

In (3.11) the matrix of the linear constraints is partitioned into the coefficient array A and the vector a_0 of the constants; A also appears in its transposed form. The objective function is decomposed into the quadratic matrix C of the coefficients of the second degree terms and into the vector c_0 of the coefficients of the linear terms. The remaining parts of the array are filled with two identity matrices and with zeroes, as shown in (3.11).

STORAGE ALLOCATION. The extra-large tableau requires some adjustments of the variables involved.

a Storage area indexed linearly from 0 to $(3 \times n + m + 2) \times (m + n + 1)$. The entities *zschrt* and *sschrt* refer to the length of the sides of the entire tableau.

proto1   Vector designating the independent variables. Length: $3 \times n + m + 1$.

proto2   Vector designating the dependent variables. Length: $m + n$.

x Solution vector. Length: $m + n$.

COMPUTATIONAL APPROACH. The method uses the long form of the Wolfe algorithm. It corresponds strictly to the description in (2.8) and the tableau a contains all variables described there. For better understanding, the meanings of the 0th row and the 0th column during the computations are given in (3.12) below:

	x^T	v^T	u^T	z^{2T}	μ	1
w	A				m	
z^1	$-2C$		A^T		n	
1	n	n	m	n	1	

$$(3.12)$$

4 ALGOL AND FORTRAN PROGRAMS

The form of the programs corresponds to the usual publication standards; in particular, all algorithms have been formulated as subroutines, and in ALGOL only lower case, in FORTRAN only upper case letters are used.

4.1 Global Procedures

(Compare: Summary 3.1; Meaning of the variables: 3.4)

4.1.1 ALGOL PROGRAMS FOR GLOBAL PROCEDURES

```
procedure   mp1 (a, znr, zschrt, sschrt, l1);
            value znr, zschrt, sschrt; integer znr, zschrt, sschrt;
            array a; integer array l1;
            comment: mp1 orders the integers in the list l1[k] such that the
            corresponding real numbers a[znr × zschrt + l1[k] × sschrt] decrease
            monotonically with k;
               begin
               integer i, k, r, s;
               for k:=2 step 1 until l1[0] do
                  begin
                  i:=0;
                  for s:=1 step 1 until k−1 do
                     if a[znr × zschrt + l1[k] × sschrt] >
                        a[znr × zschrt + l1[k − s] × sschrt] then i:=i+1;
                  if i = 0 then go to mp11;
                  r:=l1[k];
                  for s:=1 step 1 until i do l1[k−s+1]:=l1[k−s];
                  l1[k−i]:=r;
mp11:             end k;
               end mp1;
```

```
procedure   mp2 (a, l2, ip, zschrt, sschrt, kp, q1, n, v);
            value kp, zschrt, sschrt, n, v;  array a;  real q1;
            integer ip, kp, zschrt, sschrt, n, v;  integer array l2;
            comment: mp2 determines the minimum of all those numbers
            v × a[i × zschrt]/a[i × zschrt + kp × sschrt] for which a[i × zschrt]
            ≥0 and v × a[i × zschrt + kp × sschrt] >0, degeneracy is taken
            into consideration;
            begin
            integer i, io, z;  real q;
            procedure   mp4 (a, ip, kp, io, n, zschrt, sschrt, v);
                        value kp, io, n, zschrt, sschrt, v;
                        array a;  integer ip, kp, io, n, zschrt, sschrt, v;
                        comment: mp4 handles possible degeneracies;
                        begin
                        integer k;  real qp, q0;
                        for k := 1 step 1 until n do
                            begin
                            qp := v × a[ip × zschrt + k × sschrt]/a[ip ×
                                    zschrt + kp × sschrt];
                            q0 := v × a[io × zschrt + k × sschrt]/a[io ×
                                    zschrt + kp × sschrt];
                            if qp < q0 then go to mp41;
                            if qp > q0 then begin ip := io;  go to mp41
                                        end;
                            end k;
mp41:                   end mp4;
program:    ip := 0;
            for i := 1 step 1 until l2[0] do
            if v × a[l2[i] × zschrt + kp × sschrt] > 0 then
                begin
                q1 := v × a[l2[i] × zschrt]/a[l2[i] × zschrt + kp × sschrt];
                q: = q1;
                ip := l2[i];  z := i;
                go to mp21;
                end;
            go to mp22;
mp21:       for i := z + 1 step 1 until l2[0] do
                begin
                if v × a[l2[i] × zschrt + kp × sschrt] > 0 then
                q := v × a[l2[i] × zschrt]/a[l2[i] × zschrt + kp × sschrt] else
                go to mp211;
                if q < q1 then begin ip := l2[i];  q1 := q;  end;
                if q = q1 then
                    begin
                    io := l2[i];
                    mp4 (a, ip, kp, io, n, zschrt, sschrt, v);
                    end;
mp211:          end i;
mp22:       end mp2;
```

```
procedure   mp3 (a, io, i1, ko, k1, ip, kp, zschrt, sschrt, p1, p2);
            value io, i1, ko, k1, ip, kp, zschrt, sschrt, p1, p2;
            integer io, i1, ko, k1, ip, kp, zschrt, sschrt, p1, p2;
            array a;
            comment: mp3 exchanges a basic and a nonbasic variable. io
            and i1 give the rows and ko and k1 the columns to which the
            transformation is to apply. p1 and p2 indicate whether the pivotrow
            or the pivotcolumn has already been transformed;
            begin
            integer i,k; real piv;
            piv: = 1/a[ip × zschrt + kp × sschrt];
            for i: = io step 1 until ip − 1, ip + 1 step 1 until i1 do
                begin
                if p2 = 1 then
                a[i × zschrt + kp × sschrt]: = a[i × zschrt + kp × sschrt] × piv;
                for k: = ko step 1 until kp − 1, kp + 1 step 1 until k1 do
                a[i × zschrt + k × sschrt]: = a[i × zschrt + k × sschrt]
                 − a[ip × zschrt + k × sschrt] × a[i × zschrt + kp × sschrt];
                end i;
            if p1 = 1 then
            for k: = k0 step 1 until kp − 1, kp + 1 step 1 until k1 do
            a[ip × zschrt + k × sschrt]: = − a[ip × zschrt + k × sschrt] × piv;
            if p2 = 1 then a[ip × zschrt + kp × sschrt]: = piv;
            end mp3;
```

```
procedure   mp5 (a, l1, q1, kp, znr, zschrt, sschrt);
            value znr, zschrt, sschrt;  integer znr, kp, zschrt, sschrt;
            array a;  integer array l1;  real q1;
            comment: mp5 determines the minimum of the numbers
            a[znr × zschrt + l1[k] × sschrt],(k = 1, 2, . . . , l1[0]);
            begin
            integer k;
            kp: = l1[1];  q1: = a[znr × zschrt + l1[1] × sschrt];
            for k: = 2 step 1 until l1[0] do
            if a[znr × zschrt + l1[k] × sschrt] < q1 then
                begin
                q1: = a[znr × zschrt + l1[k] × sschrt];
                kp: = l1[k];
                end;
            end mp5;
```

```
procedure   mp7 (a, znr, zschrt, sschrt, kp, l1, max);
            value znr, zschrt, sschrt;  integer znr, zschrt, sschrt, kp;
            array a;  integer array l1;  real max;
            comment: mp7 determines the maximum of the numbers
            a[znr × zschrt + l1[k] × sschrt],(1 ≤ k ≤ l1[0]);
            begin
```

```
integer k;
max := a[znr × zschrt + l1[1] × sschrt];
kp := l1[1];
for k := 2 step 1 until l1[0] do
if a[znr × zschrt + l1[k] × sschrt] > max then
    begin
    max := a[znr × zschrt + l1[k] × sschrt];
    kp := l1[k];
    end;
end mp7;
```

```
procedure    mp8 (a, znr, zschrt, sschrt, list, kp, max);
             value znr, zschrt, sschrt;  integer znr, zschrt, sschrt, kp;
             real max;  array a;  integer array list;
             comment: determination of max | a[znr × zschrt + list[k] ×
             sschrt] | for 1 ≤ k ≤ list [0];
             begin
             integer k;
             kp := list[1];
             max := a[znr × zschrt + list[1] × sschrt];
             for k := 2 step 1 until list[0] do
             if abs (max) < abs (a[znr × zschrt + list[k] × sschrt]) then
                 begin
                 kp := list[k];
                 max := a[znr × zschrt + list[k] × sschrt];
                 end;
             end mp8;
```

4.1.2 ALGOL Program for the Transformatiom of the Quadratic Form in Beale's Method

(Compare: 3.5.7)

```
procedure    mp9 (a, c, ip, kp, n, m1, zschrt, sschrt, b1);
             value ip, kp, n, m1, zschrt, sschrt;
             integer ip, kp, n, m1, zschrt, sschrt;
             array a, c;  boolean b1;
             begin
             integer r, s, t, z, z1;  real store;
             comment: all comments refer to fig. 16 in 3.5.7;
program:     for r := 0 step 1 until kp − 1, kp + 1 step 1 until n do
                 begin
                 z := 0;
                 if r > kp then z1 := kp − 1;
                 if r < kp then z1 := r;
                 for s := 0 step 1 until z1 do
                     begin
                     c[r + z] := c[r + z] + a[ip × zschrt + r × sschrt] × c[kp + z];
```

```
                    z := z + n − s;
                    end s;
                t := z + kp;
                end r;
            comment: the preceding program block transforms the hatched
            elements;
mp91:       for s := kp + 1 step 1 until n do
                begin
                for r := s step 1 until n do
                c[z + n − kp + r] := c[z + n − kp + r] + a[ip × zschrt +
                    r × sschrt] × c[t + s − kp];
                z := z + n − s;
                end s;
            comment: the block from mp91 up to mp92 transforms the
            nonhatched elements;
mp92:       for r := kp + 1 step 1 until n do
            c[t + r − kp] := c[t + r − kp] + a[ip × zschrt + r × sschrt] × c[t];
            comment: between mp92 and mp93 the elements designated by 1
            are transformed;
mp93:       z1 := 0;
            for r := 0 step 1 until kp − 1 do
                begin
                z := 0;
                store := c[kp + z1] + a[ip × zschrt + r × sschrt] × c[t];
                for s := 0 step 1 until r do
                    begin
                    c[r + z] := c[r + z] + a[ip × zschrt + s × sschrt] × store;
                    z := z + n − s;
                    end s;
                z1 := z1 + n − r;
                end r;
            comment: between mp93 and mp94 the diagonally hatched
            elements contained in the upper triangle are transformed for a
            second time;
mp94:       for r := kp + 1 step 1 until n do
                begin
                z := 0;
                for s := 0 step 1 until kp − 1, kp + 1 step 1 until r do
                    begin
                    if s = kp + 1 then z := z + n − s + 1;
                    c[r + z] := c[r + z] + a[ip × zschrt + s × sschrt] ×
                        c[t + r − kp];
                    z := z + n − s;
                    end s;
                end r;
mp95:       if b1 then
                begin
                z := 0;
```

```
            for s:=0 step 1 until kp − 1 do
                begin
                c[kp + z]:= c[kp + z] + a[ip × zschrt + s × sschrt] × c[t];
                z := z + n − s;
                end s;
            for r:= kp step 1 until n do
            c[r + t − kp]:= c[r + t − kp] × a[ip × zschrt + kp × sschrt];
            z := 0;
            for s:=0 step 1 until kp do
                begin
                c[kp + z]:= c[kp + z] × a[ip × zschrt + kp × sschrt];
                z := z + n − s;
                end s;
            go to mp97;
            end;
            comment: the block between mp94 and mp95 transforms the
            elements in the hatched rectangle as well as those in the nonhatched
            triangle; in mp96 the same is done for the remaining elements
            designated by 1 and 2 provided the pivotrow is contained in the
            constraint region;
mp96:       z := 0;
            for s:=0 step 1 until kp − 1 do
                begin
                c[kp + z]:= 0;
                z := z + n − s;
                end s;
            c[t]:= 1/c[t];
            for r:= kp + 1 step 1 until n do
            c[r + t − kp]:= 0;
mp97:       end mp9;
```

4.1.3 FORTRAN Programs for the Global Procedures

```
C       SUBROUTINE MP1
C
C       MP1 ORDERS THE INTEGERS IN THE LIST L1(K) SUCH THAT THE CORRE-
C       SPONDING NUMBERS A(ZNR * ZSCHR + L1(K) * SSCHR) DECREASE MONO-
C       TONICALLY WITH K
C
        SUBROUTINE MP1(A, J1, ZNR, ZSCHR, SSCHR, L1, L10, JL1)
        INTEGER R, S, ZSCHR, SSCHR, ZNR
        DIMENSION A(J1), L1(JL1)
        IF(L10.LT.2) RETURN
        DO 1 K = 2, L10
        I = 0
        KK = K − 1
        DO 2 S = 1, KK
```

```
      KH = ZNR * ZSCHR + 1
      KH0 = KH + L1(K) * SSCHR
      KH1 = K − S
      KH = KH + L1(KH1) * SSCHR
    2 IF(A(KH0).GT.A(KH)) I = I + 1
      IF(I.EQ.0) GO TO 1
      R = L1(K)
      DO 3 S = 1, I
      KK1 = K − S
    3 L1(KK1 + 1) = L1(KK1)
      KK1 = K − 1
      L1(KK1) = R
    1 CONTINUE
      RETURN
      END
```

```
C     SUBROUTINE MP2
C
C     MP2 DETERMINES THE MINIMUM OF ALL THOSE V * A(I * ZSCHR)/
C     A(I * ZSCHR + KP * SSCHR),
C     FOR WHICH THE A(I * ZSCHR) ARE NONNEGATIVE AND THE
C     V * A(I * ZSCHR + KP * SSCHR)
C     ARE POSITIVE.
C     DEGENERACY IS TAKEN INTO CONSIDERATION.
C
      SUBROUTINE MP2 (A, J1, L2, L20, JL2, IP, ZSCHR, SSCHR, KP, Q1, N, IV)
      DIMENSION A(J1), L2(JL2)
      INTEGER ZSCHR, SSCHR, Z
      V = IV
      IP = 0
      IF(L20.LT.1) RETURN
      DO 1 I = 1, L20
      KH = L2(I) * ZSCHR + 1
      KH1 = KH + KP * SSCHR
    1 IF(V * A(KH1).GT.0.) GO TO 2
      RETURN
    2 Q1 = V * A(KH)/A(KH1)
      IP = L2(I)
      Z = I + 1
      IF(Z.GT.L20) RETURN
      DO 3 I = Z, L20
      KH = L2(I) * ZSCHR + 1
      KH1 = KH + KP * SSCHR
      IF(V * A(KH1).LE.0.) GO TO 3
      Q = V * A(KH)/A(KH1)
      IF(Q.GE.Q1) GO TO 4
      IP = L2(I)
```

```
        Q1 = Q
        GO TO 3
    4  IF(Q.NE.Q1) GO TO 3
C
C      HERE IT IS DETERMINED WHICH OF TWO ROWS WITH EQUAL
C      QUOTIENT QUALIFIES AS PIVOTROW
C
        I0 = L2(I)
        DO 5 K = 1,N
        KH0 = IP * ZSCHR + K * SSCHR + 1
        KH2 = IP * ZSCHR + KP * SSCHR + 1
        KH = I0 * ZSCHR + K * SSCHR + 1
        QP = V * A(KH0)/A(KH2)
        Q0 = V * A(KH)/A(KH1)
        IF(QP.LT.Q0) GO TO 3
    5  IF(Q0.LT.QP) GO TO 6
    6  IP = I0
    3  CONTINUE
        RETURN
        END
```

```
C      SUBROUTINE MP 3
C
C      MP3 EXCHANGES A BASIC AND A NONBASIC VARIABLE.  I0 AND I1
C      GIVE THE ROWS AND K0 AND K1 THE COLUMNS TO WHICH THE
C      TRANSFORMATION IS TO APPLY.  P1 AND P2 INDICATE WHETHER
C      THE PIVOTROW OR COLUMN HAS ALREADY BEEN TRANSFORMED.
C
        SUBROUTINE MP3 (A, J1, I0, I1, K0, K1, IP, KP, ZSCHR, SSCHR, P1, P2)
        DIMENSION A(J1)
        INTEGER ZSCHR, SSCHR, P1, P2
        KH = IP * ZSCHR + KP * SSCHR + 1
        PIV = 1./A(KH)
        II0 = I0 + 1
        II1 = I1 + 1
        KK0 = K0 + 1
        KK1 = K1 + 1
        DO 1 II = II0, II1
        I = II - 1
        IF(I.EQ.IP) GO TO 1
        KH0 = I * ZSCHR + KP * SSCHR + 1
        IF(P2.EQ.1) A(KH0) = A(KH0) * PIV
        DO 2 KK = KK0, KK1
        K = KK - 1
        IF(K.EQ.KP) GO TO 2
        KH1 = I * ZSCHR + K * SSCHR + 1
        KH2 = IP * ZSCHR + K * SSCHR + 1
        A(KH1) = A(KH1) — A(KH2) * A(KH0)
```

```
  2 CONTINUE
  1 CONTINUE
    IF(P1.NE.1) GO TO 4
    DO 5 KK = KK0, KK1
    K = KK − 1
    KH2 = IP * ZSCHR + K * SSCHR + 1
  5 IF(K.NE.KP) A(KH2) = − A(KH2) * PIV
  4 IF(P2.EQ.1) A(KH) = PIV
    RETURN
    END
```

```
  C     SUBROUTINE MP5
  C
  C     MP5 DETERMINES THE MINIMUM OF THE NUMBERS
  C     A (ZNR * ZSCHR + L1(K) * SSCHR) FOR K = 1, 2, . . . , L10.
  C     KP IS THAT INDEX K WHICH POINTS TO  THE MINIMUM Q1
  C
  C
        SUBROUTINE MP5(A, J1, L1, L10, JL1, Q1, KP, ZNR, ZSCHR, SSCHR)
        INTEGER ZNR, ZSCHR, SSCHR
        DIMENSION A(J1), L1(JL1)
        KP = L1(1)
        KH = ZNR * ZSCHR + 1
        KH0 = KH + L1(1) * SSCHR
        Q1 = A(KH0)
        IF(L10.LT.2) RETURN
        DO 1 K = 2, L10
        KH0 = KH + L1(K) * SSCHR
        IF(A(KH0).GE.Q1) GO TO 1
        Q1 = A(KH0)
        KP = L1(K)
      1 CONTINUE
        RETURN
        END
```

```
  C     SUBROUTINE MP7
  C
  C     MP7 DETERMINES THE MAXIMUM OF THOSE ELEMENTS FOR WHICH
  C     THE INDEX IS CONTAINED IN THE LIST L1
  C
        SUBROUTINE MP7(A, J1, ZNR, ZSCHR, SSCHR, KP, L1, L10, JL1, MAX)
        DIMENSION A(J1), L1(JL1)
        INTEGER ZNR, ZSCHR, SSCHR
        REAL MAX
        KH = ZNR * ZSCHR + L1(1) * SSCHR + 1
        MAX = A(KH)
        KP = L1(1)
        IF(L10.LT.2) RETURN
```

```
       DO 1 K = 2, L10
       KH = ZNR * ZSCHR + L1(K) * SSCHR + 1
       IF(A(KH).LE.MAX) GO TO 1
       MAX = A(KH)
       KP = L1(K)
     1 CONTINUE
       RETURN
       END
```

```
C      SUBROUTINE MP8
C
C      DETERMINATION OF THE MAXIMUM OF THE NUMBERS
C      ABS(A(ZNR * ZSCHR + LIST(K) * SSCHR)) FOR K = 1, 2, . . . , LIST0
C
       SUBROUTINE MP8(A, J1, ZNR, ZSCHR, SSCHR, LIST, LIST0, JLIST, KP, MAX)
       INTEGER ZNR, ZSCHR, SSCHR
       REAL MAX
       DIMENSION A(J1), LIST (JLIST)
       KP = LIST(1)
       KH = ZNR * ZSCHR + 1
       KH0 = LIST(1) * SSCHR + KH
       MAX = A(KH0)
       IF(LIST0.LT.2) RETURN
       DO 1 K = 2, LIST0
       KH0 = KH + LIST(K) * SSCHR
       IF(ABS(MAX).GE.ABS(A(KH0))) GO TO 1
       KP = LIST(K)
       MAX = A(KH0)
     1 CONTINUE
       RETURN
       END
```

```
C      MP10 IMPLEMENTS THE STEPWISE COMPUTATION OF THE INVERSE
C      MATRIX IN THE REVISED SIMPLEX METHOD
C
       SUBROUTINE MP10 (B, JB, C, JC, M, IP, S, ZSCHR1, SSCHR1)
       INTEGER ZSCHR1, SSCHR1, S
       DIMENSION B(JB), C(JC)
     1 DO 2 K = 1, S
       KH = IP * ZSCHR1 + K * SSCHR1 + 1
       IF(B(KH).EQ.0.) GO TO 2
       IM = M + 1
       DO 3 II = 1, IM
       I = II − 1
       IF(I.EQ.IP) GO TO 3
       IF(C(II).EQ.0.) GO TO 3
       KH = K * SSCHR1 + 1
```

```
      K H0 = KH + I * ZSCHR1
      KH1 = KH + IP * ZSCHR1
      B(KH0) = B(KH0) + B(KH1) * C(II)
    3 CONTINUE
    2 CONTINUE
      DO 4 K = 1, S
      KH = IP * ZSCHR1 + K * SSCHR1 + 1
    4 B(KH) = B(KH) * C(IP + 1)
      RETURN
      END
```

4.1.4 FORTRAN PROGRAM FOR THE TRANSFORMATION OF THE QUADRATIC FORM IN BEALE'S METHOD

(Compare: 3.5.7)

```
C     PROGRAM FOR THE TRANSFORMATION OF THE QUADRATIC FORM IN
C     THE METHOD OF BEALE, COMMENTS REFER TO FIGURE 16 IN SECTION
C     3.5.7
C
      SUBROUTINE MP9(A, J1, C, JC, IP, KP, N, M1, ZSCHR, SSCHR, B1)
      INTEGER ZSCHR, SSCHR, R, S, T, Z, Z1
      LOGICAL B1
      DIMENSION A(J1), C(JC)
      IN = N + 1
      DO 9000 IR = 1, IN
      R = IR - 1
      IF(R.EQ.KP) GO TO 9000
      Z = 0
      IF(R.GT.KP) Z1 = KP - 1
      IF(R.LT.KP) Z1 = R
      IZ1 = Z1 + 1
      DO 9002 IS = 1, IZ1
      S = IS - 1
      KH = R + Z + 1
      KH1 = IP * ZSCHR + R * SSCHR + 1
      KH0 = KP + Z + 1
      C(KH) = C(KH) + A(KH1) * C(KH0)
 9002 Z = Z + N - S
      T = Z + KP
 9000 CONTINUE
C
C     IN THE PROGRAM SECTION UP TO 9000 THE HATCHED ELEMENTS
C     ARE TRANSFORMED
C
 9100 IKP = KP + 1
      IF(IKP.GT.N) GO TO 9300
      DO 9102 S = IKP, N
```

```
      DO 9103 R = S, N
      KH = Z + N − KP + R + 1
      KH1 = IP * ZSCHR + R * SSCHR + 1
      KH0 = T + S − KP + 1
9103  C(KH) = C(KH) + A(KH1) * C(KH0)
9102  Z = Z + N − S
C
C     PROGRAM SECTION 91XX TRANSFORMS THE NONHATCHED PART OF
C     THE ELEMENTS
C
9200  DO 9201 R = IKP, N
      KH = T + R − KP + 1
      KH1 = IP * ZSCHR + R * SSCHR + 1
9201  C(KH) = C(KH) + A(KH1) * C(T + 1)
C
C     PROGRAM SECTION 92XX TRANSFORMS THE ELEMENTS DESIGNATED
C     BY 1
C
9300  Z1 = 0
      DO 9301 IR = 1, KP
      R = IR − 1
      Z = 0
      KH = KP + Z1 + 1
      KH1 = IP * ZSCHR + R * SSCHR + 1
      STORE = C(KH) + A(KH1) * C(T + 1)
      DO 9302 IS = 1, IR
      S = IS − 1
      KH = IP * ZSCHR + S * SSCHR + 1
      KH1 = IR + Z
      C(KH1) = C(KH1) + A(KH) * STORE
9302  Z = Z + N − S
9301  Z1 = Z1 + N − R
C
C     PROGRAM SECTION 93XX TRANSFORMS THE DIAGONALLY HATCHED
C     ELEMENTS CONTAINED IN THE UPPER TRIANGLE FOR A SECOND TIME
C
9400  IF(IKP.GT.N) GO TO 9500
      DO 9401 R = IKP, N
      Z = 0
      IR = R + 1
      DO 9401 IS = 1, IR
      S = IS − 1
      IF(S.NE.KP) GO TO 9402
      Z = Z + 1
      GO TO 9401
9402  KH = IR + Z
      KH1 = IP * ZSCHR + S * SSCHR + 1
      KH0 = T + R − KP + 1
```

```
        C(KH) = C(KH) + A(KH1) * C(KH0)
  9401  Z = Z + N - S
C
  9500  IF(.NOT.B1) GO TO 9600
        Z = 0
        KH0 = IP * ZSCHR + KP * SSCHR + 1
        DO 9501 IS = 1, KP
        S = IS - 1
        KH = IKP + Z
        KH1 = IP * ZSCHR + S * SSCHR + 1
        C(KH) = C(KH) + A(KH1) * C(T + 1)
  9501  Z = Z + N - S
        DO 9503 R = KP, N
        KH = R + T - KP + 1
  9503  C(KH) = C(KH) * A(KH0)
        Z = 0
        DO 9504 IS = 1, IKP
        S = IS - 1
        KH = IKP + Z
        C(KH) = C(KH) * A(KH0)
  9504  Z = Z + N - S
        RETURN
C
C       PROGRAM SECTIONS 94XX AND 95XX TRANSFORM THE ELEMENTS IN
C       THE HATCHED RECTANGLE, THE ELEMENTS IN THE NONHATCHED
C       TRIANGLE (UP TO 9401), AS WELL AS THE ELEMENTS DESIGNATED BY
C       1 AND 2 PROVIDED THE PIVOT ROW IS CONTAINED IN THE CONSTRAINT
C       REGION
C
  9600  Z = 0
        DO 9601 IS = 1, KP
        S = IS - 1
        KH = KP + Z + 1
        C(KH) = 0.
  9601  Z = Z + N - S
        C(T + 1) = 1./C(T + 1)
        IF(IKP.GT.N) GO TO 9602
        DO 9603 R = IKP, N
        KH = R + T - KP + 1
  9603  C(KH) = 0.
  9602  RETURN
        END
```

4.2 ALGOL Program for the Simplex Method

```
procedure    simplex (a, zschrt, sschrt, n, m1, m2, m3, fall, proto1, proto2);
             value zschrt, sschrt, n, m1, m2, m3;
             integer zschrt, sschrt, n, m1, m2, m3, fall;
             array a; integer array proto1, proto2;
             comment: routine for the optimization of a linear program using
             the simplex method, possible degeneracies are taken into considera-
             tion, simplex uses the global procedures mp2, mp3, mp7, and mp8;
             begin
             integer i, k, ip, kp, r, s, v; real max, q1;
             integer array l1[0:n], l2[0:m1 + m2 + m3], l3[0:m2];
             r:= 0; v:= − 1;
             for k:= 1 step 1 until n do l1[k]:= proto1[k]:= k;
             l1[0]:= n;
             for i:= 1 step 1 until m1 + m2 + m3 do l2[i]:= 1;
             l2[0]:= m1 + m2 + m3;
             for i:= 1 step 1 until m1 + m2 + m3 do proto2[i]:= n + i;
             if m2 + m3 = 0 then go to s3;
             comment: if the origin is a feasible solution, then the following
             block up to s3 can be bypassed;
             for i:= 1 step 1 until m2 do l3[i]:= 1;
             comment: computation of the auxiliary objective function for the
             m-method;
s00:         r:= 1;
             for k:= 0 step 1 until n do
                 begin
                 q1:= 0;
                 for i:= m1 + 1 step 1 until m1 + m2 + m3 do
                 q1:= q1 + a[i × zschrt + k × sschrt];
                 a[−zschrt + k × sschrt]:= − q1;
                 end k;
             comment: computation of a feasible solution by means of the
             simplex method, and using the above-computed auxiliary objective
             function;
s0:          mp7 (a, −1, zschrt, sschrt, kp, l1, max);
             if max ≤ 0 ∧ a[−zschrt] < 0 then
                 begin
                 fall:= 2;
                 go to s5;
                 end;
             comment: if the maximal coefficient of the auxiliary objective
             function is nonpositive and the value of the function itself is negative,
             then there exists no feasible solution;
             if max ≤ 0 ∧ a[−zschrt] = 0 then
             begin
```

```
          for ip:= m1 + m2 + 1 step 1 until m1 + m2 + m3 do
              if proto2[ip] = n + ip then
              begin mp8 (a, ip, zschrt, sschrt, l1, kp, max);
              if max > 0 then go to s01 end;
          r:= 0;
          for i:= m1 + 1 step 1 until m1 + m2 do if l3[i − m1] = 1 then
              for k:= 0 step 1 until n do
              a[i × zschrt + k × sschrt]:= − a[i × zschrt + k × sschrt];
          go to s3;
          end if;
          comment: if the above condition is satisfied, then a feasible solution
          has been found;
          mp2 (a, l2, ip, zschrt, sschrt, kp, q1, n, v);
          comment: mp2 assures that in the exchange no constraint is
          violated;
          if ip < 0 then
              begin
              fall := 2;
              go to s5;
              end;
s01:      mp3 (a, −1, m1 + m2 + m3, 0, n, ip, kp, zschrt, sschrt, 1, 1);
          if proto 2[ip] < n + m1 + m2 + 1 then go to s1;
          for k:= 1 step 1 until l1[0] do
          if l1[k] = kp then
              begin
              l1[0]:= l1[0] − 1;
              for s:= k step 1 until l1[0] do l1[s]:= l1[s + 1];
              go to s2;
              end;
s1:       if proto2[ip] < n + m1 + 1 then go to s21;
          if l3[proto2[ip] − m1 − n] = 0 then go to s21;
          l3[proto2[ip] − m1 − n]:= 0;
s2:       a[−zschrt + kp × sschrt]:= a[−zschrt + kp × sschrt] + 1;
          for i:= −1 step 1 until m1 + m2 + m3 do
          a[i × zschrt + kp × sschrt]:= − a[i × zschrt + kp × sschrt];
s21:      s:= proto1[kp];
          proto1[kp]:= proto2[ip];
          proto2[ip]:= s;
          if r ≠ 0 then go to s0;
          comment: optimization-block;
s3:       mp7 (a, 0, zschrt, sschrt, kp, l1, max);
          if max ≤ 0 then
              begin
              fall := 0;
              go to s5;
              end;
          mp2 (a, l2, ip, zschrt, sschrt, kp, q1, n, v);
```

```
s4:              if ip = 0 then
                    begin
                    fall := 1;
                    go to s5;
                    end;
                 mp3 (a, 0, m1 + m2 + m3, 0, n, ip, kp, zschrt, sschrt, 1, 1);
                 go to s21;
s5:              end simplex;
```

4.3   FORTRAN Program for the Simplex Method

```
C       ROUTINE FOR THE OPTIMIZATION OF A LINEAR PROGRAM USING THE
C       SIMPLEX METHOD.
C       DEGENERACIES ARE TAKEN INTO ACCOUNT
C       SIMPLEX USES SUBROUTINES MP2, MP3, MP7, AND MP8
C
        SUBROUTINE SIMPLEX (A, J1, ZSCHR, SSCHR, N, M1, M2, M3, FALL,
      1            PROTO1, IPROT1, PROTO2, IPROT2, L1, JL1, L2, JL2, L3, JL3)
        DIMENSION A(J1), PROTO1 (IPROT1), PROTO2 (IPROT2),
      1            L1 (JL1), L2(JL2), L3 (JL3)
        INTEGER ZSCHR, SSCHR, V, R, S, PROTO1, PROTO2, FALL
        REAL MAX
C
C       INITIALIZATION OF THE INDEX LISTS
C
        M123 = M1 + M2 + M3
        R = 0
        V = −1
        DO 1 K = 1, N
        L1(K) = K
      1 PROTO1(K) = K
        L10 = N
        DO 2 I = 1, M123
      2 L2(I) = I
        L20 = M123
        DO 3 I = 1, M123
      3 PROTO2(I) = N + I
        IF(M2 + M3.EQ.0) GO TO 103
C
C       IF THE ORIGIN IS A FEASIBLE SOLUTION, THE PROGRAM SECTION UP TO
C       103 CAN BE BYPASSED
C
        DO 17 I = 1, M2
     17 L3(I) = 1
C
```

```
C     COMPUTATION OF THE AUXILIARY OBJECTIVE FUNCTION IN PREPARA-
C     TION FOR THE M-METHOD
C
      R = 1
      N1 = N + 1
      DO 4 K = 1, N1
      KK = K - 1
      Q1 = 0.
      N2 = M1 + 1
      DO 5 I=N2, M123
      KH = I * ZSCHR + KK * SSCHR + 1
    5 Q1 = Q1 + A(KH)
      KH = (M123 + 1) * ZSCHR + KK * SSCHR + 1
    4 A(KH) = -Q1
C
C     COMPUTATION OF A FEASIBLE SOLUTION BY MEANS OF THE SIMPLEX
C     METHOD USING THE ABOVE CALCULATED AUXILIARY OBJECTIVE FUNC-
C     TION. MP7 DETERMINES THE MAXIMAL COEFFICIENT OF THE AUXILIARY
C     OBJECTIVE FUNCTION
C
  100 CALL MP7 (A, J1, M123+1, ZSCHR, SSCHR, KP, L1, L10, JL1, MAX)
      KH = (M123 + 1) * ZSCHR + 1
      IF(MAX.GT.0..OR.A(KH).GE.0.) GO TO 6
      FALL = 2
      RETURN
C
C     IF THE MAXIMAL COEFFICIENT OF THE AUXILIARY OBJECTIVE FUNCTION
C     IS NONPOSITIVE AND THE AUXILIARY OBJECTIVE FUNCTION ITSELF HAS
C     A NEGATIVE VALUE, NO FEASIBLE SOLUTION EXISTS.
C
    6 IF(MAX.GT.0..OR.A(KH).NE.0.) GO TO 16
      M12 = M1 + M2 + 1
      DO 18 IP =M12, M123
      IF(PROTO2(IP).NE.IP +N) GO TO 18
      CALL MP8(A, J1, IP, ZSCHR, SSCHR, L1, L10, JL1, KP, MAX)
      IF(MAX.GT.0.) GO TO 7
   18 CONTINUE
      R = 0
      M12 = M12 - 1
      DO 20 I = N2, M12
      KH = I - M1
      IF(L3(KH).NE.1) GO TO 20
      DO 21 K=1, N1
      KH = I * ZSCHR + (K - 1) * SSCHR + 1
   21 A(KH) = -A(KH)
   20 CONTINUE
      GO TO 103
```

```
C
C        IF THE MAXIMAL COEFFICIENT OF THE AUXILIARY OBJECTIVE FUNCTION
C        IS NONPOSITIVE AND THE VALUE OF THIS FUNCTION IS ITSELF EQUAL TO
C        ZERO, THEN A FEASIBLE SOLUTION HAS BEEN FOUND AND CONTROL
C        CAN BE TRANSFERRED TO THE OPTIMIZATION SECTION.
C
     16 CALL MP2(A, J1, L2, L20, JL2, IP, ZSCHR, SSCHR, KP, Q1, N, V)
C
C        MP2 ASSURES THAT NO CONSTRAINT IS VIOLATED IN THE EXCHANGE
C
        IF(IP.NE.0) GO TO 7
        FALL = 2
        RETURN
C
C        SINCE THE MAXIMUM OF THE AUXILIARY OBJECTIVE FUNCTION IS
C        PLUS INFINITY, NO FEASIBLE SOLUTION EXISTS
C
      7 CALL MP3(A, J1, 0, M123+1, 0, N, IP, KP, ZSCHR, SSCHR, 1, 1)
        IF(PROTO2(IP).LT.N + M1 + M2 + 1) GO TO 101
C
C        IF THE VARIABLE JUST REMOVED FROM THE BASIS BY MP3 CORRESPONDS
C        TO AN EQUALITY CONSTRAINT, THE NEXT PROGRAM SECTION ASSURES
C        THAT THIS VARIABLE REMAINS OUTSIDE THE BASIS, OTHERWISE THE
C        SECTION UP TO 101 MUST BE BYPASSED.
C
        DO 8 K = 1, L10
      8 IF(L1(K).EQ.KP) GO TO 9
      9 L10 = L10 - 1
        DO 10 S = K, L10
     10 L1(S) = L1(S + 1)
        GO TO 102
C
    101 IF(PROTO2(IP).LT.N + M1 + 1) GO TO 1021
        KH = PROTO2(IP) - M1 - N
        IF(L3(KH).EQ.0) GO TO 1021
        L3(KH) = 0
    102 KH = (M123 + 1) * ZSCHR + KP * SSCHR + 1
        A(KH) = A(KH) + 1.
        MH = M123 + 2
        DO 12 I = 1, MH
        KH = (I - 1) * ZSCHR + KP * SSCHR + 1
     12 A(KH) = -A(KH)
C
C        IN THE FOLLOWING SECTION UP TO 103 THE PROTO LISTS ARE UPDATED.
C
   1021 S = PROTO1(KP)
        PROTO1(KP) = PROTO2(IP)
```

```
      PROTO2(IP) = S
      IF(R.NE.0) GO TO 100
C
C     OPTIMIZATION SECTION
C
 103  CALL MP7(A, J1, 0, ZSCHR, SSCHR, KP, L1, L10, JL1, MAX)
      IF(MAX.GT.0.) GO TO 14
      FALL = 0
      RETURN
C
C     IF THE MAXIMAL COEFFICIENT OF THE OBJECTIVE FUNCTION IS LESS
C     THAN OR EQUAL TO ZERO, THE OPTIMAL TABLEAU HAS BEEN OBTAINED.
C
  14  CALL MP2(A, J1, L2, L20, JL2, IP, ZSCHR, SSCHR, KP, Q1, N, V)
 104  IF(IP.NE.0) GO TO 15
      FALL = 1
      RETURN
  15  CALL MP3(A, J1, 0, M123, 0, N, IP, KP, ZSCHR, SSCHR, 1, 1)
      GO TO 1021
      END
```

4.4 ALGOL Program for the Dual Simplex Method

```
procedure   dusex (a, zschrt, sschrt, n, m, fall, w, proto1, proto2);
            value zschrt, sschrt, n, m, w; integer zschrt, sschrt, n, m, fall;
            array a; integer array proto1, proto2; boolean w;
            comment: routine for determining the minimum of a linear
            program by means of the dual simplex method. the tableau is
            assumed to be dually feasible. dusex uses the global procedures mp2,
            mp3, and mp5;
            begin
            integer v, i, k, ip, kp, znr; real q1;
            integer array l1[0:m], l2[0:n];
program:    for k:=1 step 1 until m do l1[k]:=k;
            l1[0]:=m;
            for i:=1 step 1 until n do l2[i]:=i;
            l2[0]:=n;
            v:=1;
            if w then go to dusex1;
            for k:=1 step 1 until n do proto1[k]:=k;
            for i:=1 step 1 until m do proto2[i]:=n+i;
dusex1:     i:=zschrt; zschrt:=sschrt; sschrt:=i;
            i:=m; m:=n; n:=i;
            mp5(a, l1, q1, kp, 0, zschrt, sschrt);
            if q1 ≥ 0 then
```

```
            begin
            fall := 0;
            go to dusex3;
            end;
            mp2(a, l2, ip, zschrt, sschrt, kp, q1, n, v);
            if ip = 0 then
               begin
               fall := 2;
               go to dusex3;
               end;
dusex2:     i := zschrt; zschrt := sschrt; sschrt := i;
            i := m; m := n; n := i;
            i := kp; kp := ip; ip := i;
            i := proto1[kp]; proto1[kp] := proto2[ip];
            proto2[ip] := i;
            mp3(a, 0, m, 0, n, ip, kp, zschrt, sschrt, 1, 1);
            go to dusex1;
dusex3:     i := zschrt; zschrt := sschrt; sschrt := i;
            i := m; m := n; n := i;
            end dusex;
```

4.5 FORTRAN Program for the Dual Simplex Method

```
C       ROUTINE FOR DETERMINING THE MINIMUM OF A LINEAR PROGRAM
C       BY MEANS OF THE DUAL SIMPLEX METHOD. THE TABLEAU IS ASSUMED
C       TO BE DUALLY FEASIBLE. DUSEX USES THE SUBROUTINES MP2, MP3,
C       MP5.
C
        SUBROUTINE DUSEX(A, J1, ZSCHR, SSCHR, N, M, FALL, W, PROTO1,
       1              IPROT1, PROTO2, IPROT2, L1, JL1, L2, JL2)
        DIMENSION A(J1), L1(JL1), L2(JL2)
        INTEGER PROTO1 (IPROT1), PROTO2 (IPROT2), FALL, ZSCHR, SSCHR, V,
       1 ZNR
        LOGICAL W
        DO 1 K = 1, M
      1 L1(K) = K
        L10 = M
        DO 2 I = 1, N
      2 L2(I) = I
        L20 = N
        V = 1
        IF(W) GO TO 100
        DO 3 K = 1, N
      3 PROTO1(K) = K
        DO 4 I = 1, M
```

```
      4 PROTO2(I) = N + I
  C
    100 I = ZSCHR
        ZSCHR = SSCHR
        SSCHR = I
        I = M
        M = N
        N = I
        CALL MP5(A, J1, L1, L10, JL1, Q1, KP, 0, ZSCHR, SSCHR)
        IF(Q1.LT.0.) GO TO 101
        FALL = 0
        GO TO 300
    101 CALL MP2(A, J1, L2, L20, JL2, IP, ZSCHR, SSCHR, KP, Q1, N, V)
        IF(IP.NE.0) GO TO 200
        FALL = 2
        GO TO 300
  C
    200 I = ZSCHR
        ZSCHR = SSCHR
        SSCHR = I
        I = M
        M = N
        N = I
        I = KP
        KP = IP
        IP = I
        I = PROTO1(KP)
        PROTO1(KP) = PROTO2(IP)
        PROTO2(IP) = I
        CALL MP3(A, J1, 0, M, 0, N, IP, KP, ZSCHR, SSCHR, 1, 1)
        GO TO 100
  C
    300 I = ZSCHR
        ZSCHR = SSCHR
        SSCHR = I
        I = M
        M = N
        N = I
        RETURN
        END
```

4.6 ALGOL Program for the Revised Simplex Method

```
procedure    resex (a, n, m, zschrt, sschrt, fall, proto1, proto2);
             value n, m, zschrt, sschrt; integer n, m, zschrt, sschrt, fall;
             array a; integer array proto1, proto2;
             comment: routine for computing the maximum of a linear program
             by means of the revised simplex method. resex presupposes that the
             origin is a feasible solution. resex uses the global procedure mp7;
             begin
             integer i, k, ip, kp, zschrt1, sschrt1, s, j;
             real max, q1; array b[m + 1 :m × m + m + m], c[0 :m];
             integer array l1[0 :n], l2, l3[0 :m];
             procedure mp10(b1, c1, m, ip, s, zschrt2, sschrt2);
                        value m, s, ip, zschrt2, sschrt2;
                        integer m, s, ip, zschrt2, sschrt2; array b1, c1;
                        comment: routine for the stepwise computation of
                        the inverse matrix occurring in the revised simplex
                        method;
                        begin
                        integer i, k;
                        for k := 1 step 1 until s do
                        if b1[ip × zschrt2 + k × sschrt2] ≠ 0 then
                        for i := 0 step 1 until ip − 1, ip + 1 step 1 until
                        m do
                        if c1[i] ≠ 0 then
                        b1[i × zschrt2 + k × sschrt2] :=
                                      b1[i × zschrt2 + k × sschrt2] +
                                      b1[ip × zschrt2 + k × sschrt2] ×
                                      c1[i];
                        for k := 1 step 1 until s do
                        b1[ip × zschrt2 + k × sschrt2] :=
                                      b1[ip × zschrt2 + k × sschrt2] × c1[ip];
                        end mp10;
program:     for k := 1 step 1 until n do proto1[k] := l1[k] := k;
             for i := 1 step 1 until m do proto2[i] := n + i;
             l1[0] := n; l2[0] := 0;
             zschrt1 := 1; sschrt1 := m + 1;
             comment: generation of the identity matrix;
             for i := 0 step 1 until m do
             for k := 1 step 1 until m do
             b[i × zschrt1 + k × sschrt1] := if k = i then 1 else 0;
             comment: determination of the pivot-column
r1:          max := 0; q1 := 0;
             if l1[0] = 0 then go to r2;
             mp7 (a, 0, zschrt, sschrt, kp, l1, max);
r2:          if l2[0] = 0 then go to r3;
             for i := 1 step 1 until l2[0] do
```

```
        if b[0 × zschrt1 + l2[i] × sschrt1] > q1 then
           begin
           s := i;
           q1 := b[0 × zschrt1 + l2[i] × sschrt1];
           end;
r3:     if max ≤ 0 ∧ q1 ≤ 0 then
           begin fall := 0; go to r9 end;
        k := 0;
        if max ≥ q1 then go to r4;
        k := 1; kp := l3[s];
        for i := 0 step 1 until m do
        c[i] := b[i × zschrt1 + s × sschrt1];
        go to r5;
r4:     c[0] := a[0 × zschrt + kp × sschrt];
        for i := 1 step 1 until m do
           begin
           q1 := 0;
           for j = 1 step 1 until m do
           if a[j × zschrt + kp × sschrt] ≠ 0 ∧ b[i × zschrt1
              + j × sschrt1] ≠ 0 then
           q1 := q1 + a[j × zschrt + kp × sschrt]
              × b[i × zschrt1 + j × sschrt1];
           c[i] := q1;
           end i;
        comment: determination of the pivotrow;
r5:     ip := 0;
        for i := 1 step 1 until m do
        if c[i] > 0 then
           begin
           q1 := a[i × zschrt]/c[i];
           ip := i;
           go to r6;
           end;
        fall := 1;
        go to r9;
r6:     j := ip;
        for i := j step 1 until m do
        if c[i] > 0 then
           begin
           if a[i × zschrt]/c[i] < q1 then
              begin q1 := a[i × zschrt]/c[i]; ip := i; end;
           end;
        for i := 0 step 1 until ip − 1, ip + 1 step 1 until m do
        c[i] := −c[i]/c[ip];
        c[ip] := 1/c[ip];
        comment: transformation of the first column of tableau a;
        mp10(a, c, m, ip, 1, zschrt, 0);
```

```
            comment: transformation of the first row of tableau a;
            for j := 1 step 1 until kp − 1, kp + 1 step 1 until n do
                begin
                q1 := 0;
                for i := 1 step 1 until m do
                if b[ip × zschrt1 + i × sschrt1] ≠ 0 then
                q1 := q1 + b[ip × zschrt1 + i × sschrt1]
                        × a[i × zschrt + j × sschrt];
                a[0 × zschrt + j × sschrt] := a[0 × zschrt + j × sschrt]
                        − q1 × c[0];
                end j;
            a[0 × zschrt + kp × sschrt] := c[0];
            comment: transformation of the inverse;
            mp10(b, c, m, ip, m, zschrt1, sschrt1);
            if k = 0 then
                begin
                for j := 1 step 1 until l2[0] do
                if ip = l2[j] then
                    begin
                    if l3[j] = kp then go to r8;
                    for i := 1 step 1 until l1[0] do
                    if l1[i] = kp then l1[i] := l3[j];
                    l3[j] := kp;
                    go to r8;
                    end;
                l2[0] := l2[0] + 1; l2[l2[0]] := ip; l3[l2[0]] := kp;
r7:             for i := 1 step 1 until l1[0] do
                if l1[i] = kp then
                    begin
                    l1[0] := l1[0] − 1;
                    for j := i step 1 until l1[0] do
                    l1[j] := l1[j + 1];
                    go to r8;
                    end;
                end;
            for j := 1 step 1 until l2[0] do
            if ip = l2[j] then go to r8;
            l1[0] := l1[0] + 1; l1[l1[0]] := l3[s]; l2[0] := l2[0] − 1;
            for j := s step 1 until l2[0] do
                begin l2[j] := l2[j + 1]; l3[j] := l3[j + 1]; end;
r8:         k := proto1[kp]; proto1[kp] := proto2[ip]; proto2[ip] := k;
            go to r1;
r9:         end resex;
```

4.7 FORTRAN Program for the Revised Simplex Method

```
C      ROUTINE FOR DETERMINING THE MAXIMUM OF A LINEAR PROGRAM
C      BY MEANS OF THE SIMPLEX METHOD. RESEX PRESUPPOSES THAT THE
C      ORIGIN IS A FEASIBLE SOLUTION.  RESEX USES THE SUBROUTINES MP7
C      AND MP10.
C
       SUBROUTINE RESEX(A, J1, B, JB, C, JC, N, M, ZSCHR, SSCHR, FALL, PROTO1,
      1              IPROT1, PROTO2, IPROT2, L1, JL1, L2, JL2, L3, JL3)
       DIMENSION A(J1), B(JB), C(JC), L1(JL1), L2(JL2), L3(JL3)
       INTEGER PROTO1(IPROT1), PROTO2(IPROT2), FALL, ZSCHR, SSCHR, S,
      1              ZSCHR1, SSCHR1
       REAL MAX
       DO 1 K = 1, N
       L1(K) = K
     1 PROTO1(K) = K
       DO 2 I = 1, M
     2 PROTO2(I) = N + I
       L10 = N
       L20 = 0
       ZSCHR1 = 1
       SSCHR1 = M + 1
C
C      GENERATION OF THE IDENTITY MATRIX
C
       M1 = M + 1
       DO 3 I = 1, M1
       DO 3 K = 1, M
       KH = (I − 1) * ZSCHR1 + K * SSCHR1 + 1
       B(KH) = 0.
     3 IF(K.EQ.I − 1) B(KH) = 1.
C
C      DETERMINATION OF THE PIVOT COLUMN
C
   100 MAX = 0
       Q1 = 0
       IF(L10.EQ.0) GO TO 200
       CALL MP7(A, J1, 0, ZSCHR, SSCHR, KP, L1, L10, JL1, MAX)
   200 IF(L20.EQ.0) GO TO 300
       DO 201 I = 1, L20
       KH = L2(I) * SSCHR1 + 1
       IF(B(KH).LE.Q1) GO TO 201
       S = I
       Q1 = B(KH)
   201 CONTINUE
```

```
300 IF(MAX.GT.0..OR.Q1.GT.0.) GO TO 301
    FALL = 0
    RETURN
301 K = 0
    IF(MAX.GE.Q1) GO TO 400
    K = 1
    KP = L3(S)
    DO 302 II = 1, M1
    I = II − 1
    KH = I * ZSCHR1 + S * SSCHR1 + 1
302 C(II) = B(KH)
    GO TO 500
400 KH = KP * SSCHR + 1
    C(1) = A(KH)
    DO 401 I = 1, M
    Q1 = 0.
    DO 402 J = 1, M
    KH1 = KH + J * ZSCHR
    KH2 = I * ZSCHR1 + J * SSCHR1 + 1
402 IF(A(KH1).NE.0..AND.B(KH2).NE.0.) Q1 = Q1 + A(KH1) * B(KH2)
401 C(I + 1) = Q1
C
C
C       DETERMINATION OF THE PIVOTROW
C
500 IP = 0
    DO 501 I = 1, M
501 IF(C(I + 1).GT.0.) GO TO 502
    FALL = 1
    RETURN
502 KH = I * ZSCHR + 1
    Q1 = A(KH)/C(I + 1)
    IP = I
600 J = IP
    DO 601 I = J, M
    IF(C(I + 1).LE.0.) GO TO 601
    KH = I * ZSCHR + 1
    IF(A(KH)/C(I + 1).GE.Q1) GO TO 601
    Q1 = A(KH)/C(I + 1)
    IP = I
601 CONTINUE
    C(IP + 1) = 1./C(IP + 1)
    DO 602 II = 1, M1
602 IF(II − 1.NE.IP) C(II) = −C(II) * C(IP + 1)
C
C       TRANSFORMATION OF THE FIRST COLUMN OF TABLEAU A
C
    CALL MP10(A, J1, C, JC, M, IP, 1, ZSCHR, 0)
```

```
C
C       TRANSFORMATION OF THE FIRST ROW OF TABLEAU A
C
        DO 603 J = 1, N
        IF(J.EQ.KP) GO TO 603
        Q1 = 0.
        DO 604 I = 1, M
        KH = IP * ZSCHR1 + I * SSCHR1 + 1
        KH1 = I * ZSCHR + J * SSCHR + 1
    604 IF(B(KH).NE.0.) Q1 = Q1 + B(KH) * A(KH1)
        KH = J * SSCHR + 1
        A(KH) = A(KH) − Q1 * C(1)
    603 CONTINUE
        KH = KP * SSCHR + 1
        A(KH) = −C(1)
C
C       TRANSFORMATION OF THE INVERSE
C
        CALL MP10(B, JB, C, JC, M, IP, M, ZSCHR1, SSCHR1)
        IF(K.NE.0) GO TO 701
        IF(L20.LT.1) GO TO 605
        DO 606 J = 1, L20
    606 IF(IP.EQ.L2(J)) GO TO 607
        GO TO 605
    607 IF(L3(J).EQ.KP) GO TO 800
        IF(L10.LT.1) GO TO 609
        DO 608 I = 1, L10
    608 IF(L1(I).EQ.KP) L1(I) = L3(J)
    609 L3(J) = KP
        GO TO 800
    605 L20 = L20 + 1
        L2(L20) = IP
        L3(L20) = KP
    700 IF(L10.LT.1) GO TO 800
        DO 702 I = 1, L10
    702 IF(L1(I).EQ.KP) GO TO 704
        GO TO 800
    704 L10 = L10 − 1
        DO 703 J = I, L10
    703 L1(J) = L1(J + 1)
        GO TO 800
    701 IF(L20.LT.1) GO TO 707
        DO 705 J = 1, L20
    705 IF(IP.EQ.L2(J)) GO TO 800
    707 L10 = L10 + 1
        L1(L10) = L3(S)
```

```
        L20 = L20 − 1
        DO 706 J = S, L20
        L2(J) = L2(J + 1)
    706 L3(J) = L3(J + 1)
    800 K = PROTO1(KP)
        PROTO1(KP) = PROTO2(IP)
        PROTO2(IP) = K
        GO TO 100
        END
```

4.8 ALGOL Program for the Decomposition Algorithm

procedure *decomp (a, b, x, n, m, i1, i2, i3, i4, zschrt, sschrt, list1, list2, list3*
list4, list5, proto, fall);
value *n, m, i1, i2, i3, i4, zschrt, sschrt;*
integer *n, m, i1, i2, i3, i4, zschrt, sschrt, fall;*
array *a, b, x;*
integer array *list1, list2, list3, list4, list5, proto;*
comment: *routine for computing the maximum of a linear program*
by means of the decomposition algorithm. decomp uses the global
procedures pricevector and up;
begin
array $p[1:m + list2[0]]$, $c[1:(m + list2[0]) \times i3]$,
$d[1:i2]$, $la[1:m + list2[0]]$;
integer array $l1[0:m + list2[0]]$, $proto1[1:i3]$, $proto2[1:i4]$;
integer *i, k, r, s, ip, kp, zschrt1, sschrt1;* **real** *q, max;*

program: **for** $i := 1$ **step** 1 **until** $m + list2[0]$ **do**
begin $p[i] := 0$; $l1[i] := 0$; **end;**
$l1[0] := 0$;
for $i := 1$ **step** 1 **until** n **do** $proto[i] := 0$;
d0: $max := 0$; $r := 0$;
for $k := 1$ **step** 1 **until** n **do**
    begin
    for $i := 1$ **step** 1 **until** $list1[k]$ **do**
        begin
        $q := 0$;
        for $s := 1$ **step** 1 **until** m **do**
        $q := q + p[s] \times a[s \times zschrt + (r + i) \times sschrt];$
        $d[r + i] := b[list5[k] + 0 \times list3[k] + i \times list4[k]]$
                $:= a[0 \times zschrt + (r + i) \times sschrt] − q;$
        end *i;*
    up (b, list1[k], list2[k], list3[k], list4[k], list5[k],
        proto1, proto2, fall);
    if *fall* $\neq 0$ **then go to** *d4;*
```
```

```
              for i:= 1 step 1 until list1[k] do x[i]:= 0;
              for i:= 1 step 1 until list2[k] do
              if proto2[i] ≤ list1[k] then
              x[proto2[i]]:= b[list5[k] + i × list3[k]];
              q:= 0;
              for i:= 1 step 1 until list1[k] do
              q:= q + x[i] × d[r + i];
              q:= q − p[m + k];
              r:= r + l1[k];
              if q > max then begin max:= q; kp:= k; end;
              end k;
              if max = 0 then go to d2;
              pricevector (p, c, la, ip, kp, a, x, l1, proto, zschrt1, sschrt1);
              if ip = 0 then begin fall:= 1; go to d4; end;
d2:           r:= kp:= fall:= 0;
              for i:= 0 step 1 until i2 do x[i]:= 0;
              for k:= 1 step 1 until n do
                 begin
                 if proto[k] = 0 then go to d3;
                 for i:= 1 step 1 until list1[k] do
                 for s:= 1 step 1 until proto[k] do
                 x[i + r]:= la[s + kp] × c[i × zschr1 + (kp + s) × sschrt1] +
                       x[i + r];
                 kp:= kp + proto[k];
d3:              r:= r + l1[k];
                 end k;
              for i:= 1 step 1 until i2 do
              x[0]:= x[0] + x[i] × a[0 × zschrt + i × sschrt];
d4:           end decomp;
```

4.9   FORTRAN Program for the Decomposition Algorithm

```
C     ROUTINE FOR COMPUTING THE MAXIMUM OF A LINEAR PROGRAM BY
C     MEANS OF THE DECOMPOSITION ALGORITHM.  DECOMP USES THE
C     SUBROUTINES PRSVKT AND UP
C     PRSVKT CORRESPONDS ESSENTIALLY TO ONE STEP OF THE REVISED
C     SIMPLEX METHOD.  UP CAN BE REPLACED, FOR EXAMPLE, BY SIMPLEX,
C     RESEX OR A ROUTINE FOR A LINEAR TRANSPORTATION METHOD.
C
      SUBROUTINE DECOMP (A, NA, B, NB, X, NX, N, M, I1, I2, I3, I4, ZSCHR, SSCHR,
     1              LIST1, NLIST1, LIST2, NLIST2, LIST3, NLIST3, LIST4,
     2              NLIST4, LIST5, NLIST5, LIST20, PROTO, NPROTO, FALL,
     3              LA, NLA, D, ND, PROTO1, NPROT1, PROTO2, NPROT2, P,
     4              NP, C, NC, L, NL, UP)
      DIMENSION LIST1(NLIST1), LIST2(NLIST2), LIST3(NLIST3), LIST4(NLIST4),
     1      LIST5(NLIST5), A(NA), B(NB), X(NX), L(NL), P(NP), C(NC), D(ND)
      INTEGER R, S, ZSCHR, SSCHR, ZSCHR1, SSCHR1, PROTO(NPROTO), PROTO1
```

```
    1     (NPROT1), PROTO2(NPROT2), FALL
          REAL LA(NLA), MAX
C         EXTERNAL UP
          M1 = M + LIST20
          DO 1 I = 1, M1
          P(I) = 0.
    1 L(I) = 0
          LO = 0
          DO 2 I = 1, N
    2 PROTO(I) = 0
          MAX = 0.
          R = 0
          DO 3 K = 1, N
          M1 = LIST1(K)
          DO 4 I = 1, M1
          Q = 0.
          DO 5 S = 1, M
          KH = S * ZSCHR + (R + I) * SSCHR + 1
    5 Q = Q + P(S) * A(KH)
          KH = R + 1
          KH1 = LIST5(K) + I * LIST4(K) + 1
          KH2 = KH * SSCHR + 1
          D(KH) = A(KH2) − Q
    4 B(KH1) = D(KH)
          CALL UP(B, NB, LIST1(K), LIST2(K), LIST3(K), LIST4(K), LIST5(K), PROTO1,
    1     NPROT1, PROTO2, NPROT2, FALL)
          IF(FALL.NE.0) GO TO 400
          DO 6 I = 1, M1
    6 X(I + 1) = 0.
          M2 = LIST2(K)
          DO 7 I = 1, M2
          KH = PROTO2(I)
          KH1 = LIST5(K) + I * LIST3(K) + 1
    7 IF (KH.LE.M1) X (KH + 1) = B(KH1)
          Q = 0.
          DO 8 I = 1, M1
          KH = R + 1
    8 Q = Q + X(I + 1) * D(KH)
          KH = M + K
          Q = Q − P(KH)
          R = R + L(K)
          IF(Q.LE.MAX) GO TO 3
          MAX = Q
          KP = K
    3 CONTINUE
          IF(MAX.EQ.0.) GO TO 200
          CALL PRSVKT (P, NP, C, NC, LA, NLA, IP, KP, A, NA, X, NX, L, NL, PROTO,
    1          NPROTO, ZSCHR1, SSCHR1)
```

```
      IF(IP.NE.0) GO TO 200
      FALL = 1
      GO TO 400
200   R = 0
      KP = 0
      FALL = 0
      M1 = I2 + 1
      DO 201 I = 1, M1
201   X(I) = 0.
      DO 202 K = 1, N
      IF(PROTO(K).EQ.0) GO TO 202
      M2 = LIST1(K)
      DO 203 I = 1, M2
      M3 = PROTO(K)
      DO 203 S = 1, M3
      KH = I + R + 1
      KH1 = S + KP
      KH2 = I * ZSCHR1 + (KP + S) * SSCHR1
203   X(KH) = X(KH) + C(KH2) * LA(KH1)
      KP = KP + PROTO(K)
202   R = R + L (K)
      DO 204 I = 1, I2
      KH = I * SSCHR + 1
204   X(1) = X(1) + X(I + 1) * A(KH)
400   RETURN
      END
```

4.10 ALGOL Program for the Duoplex Method

procedure *duoplex (a, n, m1, m2, zschrt, sschrt, fall, proto1, proto2);*
          value *n, m1, m2, zschrt, sschrt;*
          integer *n, m1, m2, zschrt, sschrt, fall;*
          array *a;* **integer array** *proto1, proto2;*
          comment: *routine to determine the maximum of a linear program by means of the duoplex method. duoplex uses the global procedures mp1, mp2, mp3, and mp5;*
          begin
          integer *v, i, k, ip, kp, r, s, z, j;* **real** *q1, q2, q3;*
          integer array *I1[0:n], I2[0:m1 + m2], I3[0:m2];*
          procedure *list (m1, m2, a, I2, z, zschrt);*
               value *m1, m2, zschrt;*
               integer *m1, m2, z, zschrt;*
               integer array *I2;* **array** *a;*
               comment: *list inserts all row indices i with $a[i \times zschrt] \geq 0$, $i \geq 1$, into list I2 and associates z with that index of the first row*

```
                    for which a[i × zschrt] < 0;
                    begin
                    integer i;
                    l2[0]:= z := 0;
                    for i := 1 step 1 until m1 + m2 do
                        begin
                        if a[i × zschrt] ≥ 0 then
                            begin
                            l2[0]:= l2[0] + 1;
                            l2[l2[0]]:= i;
                            go to list1;
                            end;
                        if z = 0 then z := i;
                        end i;
list1:              end list;
program:            l1[0]:= n;  j := 1;  r := 0;
                    for k := 1 step 1 until n do l1[k]:= proto1[k]:= k;
                    for i := 1 step 1 until m1 + m2 do proto2[i]:= n + i;
                    if m2 = 0 then begin l3[0]:= 0;  go to mw;  end;
                    comment: if there are no equality constraints, the next block up
                    to mw can be bypassed;
equations:          l3[0]:= m2;
                    for i := 1 step 1 until m2 do l3[i]:= m1 + i;
                    comment: variables belonging to the equations are removed from
                    the basis by means of the multiphase method;
gl1:                list (m1, m2, a, l2, z, zschrt);
                    mp5 (a, l1, q1, kp, l3[1], zschrt, sschrt);
                    if q1 ≥ 0 then
                        begin
                        fall := 2;
                        go to duoplex1;
                        end;
                    mp2 (a, l2, ip, zschrt, sschrt, kp, q1, n, −1);
gl2:                if ip = 0 then
                        begin
                        ip := l3[1];
                        l3[0]:= l3[0] − 1;
                        for s := 1 step 1 until l3[0] do
                            l3[s]:= l3[s + 1];
                        l1[0]:= l1[0] − 1;
                        for s := 1 step 1 until l3[0] do
                            l1[m1 + s]:= l1[m1 + s + 1];
                        go to gl3;
                        end;
                    q2 := a[l3[1] × zschrt]/(−a[l3[1] × zschrt + kp × sschrt]);
                    if q2 ≤ q1 then
                        begin
                        ip := 0;
```

```
            go to gl2;
            end;
        if ip ≤ m1 then go to gl3;
        for s:= 1 step 1 until l3[0] do
        if ip = l3[s] then
            begin
            l3[0]:= l3[0] − 1;  l1[0]:= l1[0] − 1;
            for r:= s step 1 until l3[0] do
            begin l3[r]:= l3[r + 1];  l1[m1 + r]:= l1[m1 + r + 1] end;
            end;
gl3:        s:= proto1[kp];
            proto1[kp]:= proto2[ip];
            proto2[ip]:= s;
            mp3(a, 0, m1 + m2, 0, n, ip, kp, zschrt, sschrt, 1, 1);
            if l3[0] = 0 then go to mw;
            go to gl1;
            comment: in the block beginning with mw that constraint is
            determined for which the angle between its normal and the gradient
            of the objective function is largest;
mw:         list (m1, m2, a, l2, z, zschrt);
            mp1 (a, 0, zschrt, sschrt, l1);
            kp:= l1[1];
            if a[l1[1] × sschrt] ≤ 0 ∧ z = 0 then
                begin
                if r = 0 ∨ a[r × sschrt] ≤ 0 then
                    begin
                    fall:= 0;
                    go to duoplex1;
                    end;
                l1[0]:= l1[0] + 1;
                l1[l1[0]]:= r;
                kp:= r;  r:= 0;
                go to s2;
                end;
            if j = 0 then go to s1;
            q2:= ₁₀30;
            for i:= 1 step 1 until m1 + m2 do
                begin
                q1:= q2:= 0;
                for k:= 1 step 1 until l1[0] do
                    begin
                    q1:= q1 + a[l1[k] × sschrt] × a[i × zschrt + l1[k] × sschrt]
                    q3:= q3 + a[i × zschrt + l1[k] × sschrt] ↑ 2;
                    end k;
                if q1 ≥ 0 ∧ q2 < 0 then go to mw1;
                if q1 < 0 then q1:= − q1 ↑ 2 else q1:= q1 ↑ 2;
                q1:= q1/q3;
                if q1 < q2 then begin q2:= q1;  ip:= i;  end;
```

```
mw1 :           end i;
                k := 1;
mw2 :           if a[ip × zschrt + l1[k] × sschrt] ≠ 0 then
                    begin
                    r := kp := l1[k];
                    l1[0] := l1[0] − 1;
                    for s := k step 1 until l1[0] do l1[s] := l1[s + 1];
                    j := 0;
                    go to gl3;
                    end;
                k := k + 1;
                go to mw2;
                comment: in the next block the still violated constraints are
                being satisfied;
s1 :            if z = 0 then go to s2;
                mp1 (a, z, zschrt, sschrt, l1);
                if a[z × zschrt + l1[1] × sschrt] ≤ 0 then
                    begin
                    if a[z × zschrt + r × sschrt] ≤ 0 then
                        begin
                        fall := 2;
                        go to duoplex1 :
                        end;
                    kp := r;
                    l1[0] := l1[0] + 1;
                    l1[l1[0]] := r;
                    r := 0;
                    go to s2;
                    end;
                kp := l1[1];
s2 :            mp2 (a, l2, ip, zschrt, sschrt, kp, q1, n, −1);
                if ip = 0 then
                    begin
                    if z = 0 then
                        begin
                        fall := 1;
                        go to duoplex1;
                        end;
                    ip := z;
                    end;
                go to gl3;
duoplex1 :      end duoplex;
```

4.11   FORTRAN Program for the Duoplex Method

```
C     ROUTINE FOR DETERMINING THE MAXIMUM OF A LINEAR PROGRAM.
C     DUOPLEX USES THE SUBROUTINES MP1, MP2, MP3, AND MP5.
```

```
C
      SUBROUTINE DUOPLX(A, J1, N, M1, M2, ZSCHR, SSCHR, FALL, PROTO1,
     1     IPROT1, PROTO2, IPROT2, L1, JL1, L2, JL2, L3, JL3)
      DIMENSION A(J1), PROTO1(IPROT1), PROTO2(IPROT2), L1(JL1), L2(JL2),
     1     L3(JL3)
      INTEGER PROTO1, PROTO2, ZSCHR, SSCHR, FALL, R, S, Z
C
C     IN THE FOLLOWING SECTION THE INDEX LISTS ARE INITIALIZED.
C
      L10 = N
      J = 1
      DO 1 K = 1, N
      L1(K) = K
    1 PROTO1(K) = K
      M12 = M1 + M2
      DO 2 I = 1, M12
    2 PROTO2(I) = N + I
      IF(M2.NE.0) GO TO 100
      L30 = 0
      GO TO 300
C
C     IF THERE ARE NO EQUALITY-CONSTRAINTS, THE FOLLOWING SECTION
C     UP TO STATEMENT 300 CAN BE BYPASSED.
C
  100 L30 = M2
      DO 101 I = 1, M2
  101 L3(I) = M1 + I
C
C     IN THE SECTION WITH LABELS 1XX THE BASIC VARIABLES BELONGING
C     TO EQUATIONS ARE REMOVED FROM THE BASIS. THIS IS ACCOMPLISHED
C     BY MEANS OF THE MULTIPHASE METHOD.
C
 1000 ASSIGN 102 TO KW
      GO TO 500
C
C     EXECUTION OF PROGRAM SECTION LIST·
C
  102 CALL MP5(A, J1, L1, L10, JL1, Q1, KP, L3(1), ZSCHR, SSCHR)
      IF(Q1.LT.0.) GO TO 103
      FALL = 2
      RETURN
C
C     IF ALL COEFFICIENTS OF AN EQUALITY-CONSTRAINT ARE NONNEGATIVE,
C     NO FEASIBLE SOLUTION EXISTS.
C
  103 CALL MP2(A, J1, L2, L20, JL2, IP, ZSCHR, SSCHR, KP, Q1, N, −1)
  120 IF(IP.NE.0) GO TO 121
      IP = L3(1)
```

```
          L30 = L30 − 1
          IF(L30.LT.1) GO TO 122
          DO 123 S = 1, L30
          KH = M1 + S
          L1(KH) = L1(KH + 1)
      123 L3(S) = L3(S + 1)
      122 L10 = L10 − 1
          GO TO 130
      121 KH = L3(1) ∗ ZSCHR + 1
          KH1 = KH + KP ∗ SSCHR
          Q2 = A(KH)/(−A(KH1))
          IF(Q2.GT.Q1) GO TO 124
          IP = 0
          GO TO 120
C
C         IP.EQ.0 MEANS THAT NO PRIOR SATISFIED CONSTRAINT CAN BE VIO-
C         LATED AND THAT THE PARTICULAR CONSTRAINT CAN BE SATISFIED
C         IN ONLY ONE EXCHANGE STEP. THIS IS ALSO POSSIBLE IF Q2 DOES
C         NOT EXCEED Q1.
C
      124 IF(IP.LE.M1) GO TO 130
          IF(L30.LT.1) GO TO 130
          DO 125 S = 1, L30
      125 IF(IP.EQ.L3(S)) GO TO 126
          GO TO 130
      126 I = L3(S)
          II = S − 1
          IF(II.LT.1) GO TO 128
          DO 127 R = 1, II
          KH = S − R
      127 L3(KH + 1) = L3(KH)
      128 L3(1) = I
          KH = M1 + S
          I = L1(KH)
          IF(II.LT.1) GO TO 129
          DO 1290 R = 1, II
          KH = M1 + S − R
     1290 L1(KH + 1) = L1(KH)
      129 L1(M1 + 1) = I
          GO TO 120
C
      130 S = PROTO1(KP)
          PROTO1(KP) = PROTO2(IP)
          PROTO2(IP) = S
          CALL MP3(A, J1, 0, M12, 0, N, IP, KP, ZSCHR, SSCHR, 1, 1)
          IF(L30.NE.0) GO TO 1000
C
C         THE FOLLOWING SECTION DETERMINES THE CONSTRAINT FOR WHICH
C         THE ANGLE BETWEEN ITS NORMAL AND THE GRADIENT OF THE
```

```
C      OBJECTIVE FUNCTION IS LARGEST.
C
  300 ASSIGN 301 TO KW
      GO TO 500
  301 CALL MP1(A, J1, 0, ZSCHR, SSCHR, L1, L10, JL1)
      KP = L1(1)
      KH = L1(1) * SSCHR + 1
      IF(A(KH).GT.0..OR.Z.NE.0) GO TO 302
      FALL = 0
      RETURN
C
C      THE PRECEDING SECTION CONTAINS THE OPTIMALITY CRITERION
C
  302 IF(J.EQ.0) GO TO 410
C
C      THE SECTION FROM 300 UP TO THIS POINT IS ALSO USED FOR ALL
C      SUBSEQUENT ITERATION STEPS. THE NEXT SECTION UP TO 410
C      CONTAINS THE ACTUAL DETERMINATION OF THE MAXIMAL ANGLE.
C
      Q2 = 1.E + 30
      DO 303 I = 1, M12
      Q1 = 0.
      Q3 = 0.
      IF(L10.LT.1) GO TO 305
      DO 304 K = 1, L10
      KH = L1(K) * SSCHR + 1
      KH1 = I * ZSCHR + KH
      Q1 = Q1 + A(KH) * A(KH)
  304 Q3 = Q3 + A(KH1) * A(KH1)
  305 IF(Q1.GE.0..AND.Q2.LT.0.) GO TO 303
      Q1 = SIGN(Q1 * Q1, Q1)/Q3
      IF(Q1.GE.Q2) GO TO 303
      Q2 = Q1
      IP = I
  303 CONTINUE
C
C      THE MINIMUM JUST NOW COMPUTED DETERMINES THE PIVOT-
C      COLUMN. THE PIVOTCOLUMN IS DETERMINED BY THAT MAXIMAL
C      COEFFICIENT OF THE OBJECTIVE FUNCTION WHICH CORRESPONDS
C      TO A NONZERO PIVOTELEMENT.
C
      K = 1
  306 KH = IP * ZSCHR + L1(K) * SSCHR + 1
      IF(A(KH).EQ.0.) GO TO 307
      R = L1(K)
      KP = R
      L10 = L10 - 1
      IF(K.GT.L10) GO TO 309
```

```
          DO 308 S = K, L10
     308  L1(S) = L1(S + 1)
     309  J = 0
          GO TO 130
     307  K = K + 1
          GO TO 306
C
C         IN THE FOLLOWING SECTION THE VIOLATED CONSTRAINTS ARE
C         SATISFIED AND (IF POSSIBLE) THE VALUE OF THE OBJECTIVE FUNCTION
C         IS INCREASED BY MEANS OF THE SIMPLEX METHOD.
C
     410  IF(Z.EQ.0) GO TO 420
          CALL MP1(A, J1, Z, ZSCHR, SSCHR, L1, L10, JL1)
          KP = L1(1)
          CH = Z * ZSCHR + L1(1) * SSCHR + 1
          IF(A(KH).GT.0.) GO TO 420
          KH = Z * ZSCHR + R * SSCHR + 1
          IF(A(KH).GT.0.) GO TO 411
          FALL = 2
          RETURN
     411  KP = R
          L10 = L10 + 1
          L1(L10) = R
          R = 0
C
C         THE COLUMN DETERMINED BY THE DUOPLEX METHOD PLAYS A
C         SPECIAL ROLE, IT IS CHOSEN AS THE PIVOTCOLUMN (WHENEVER
C         POSSIBLE) ONLY IF NO OTHER PIVOTCOLUMN CAN BE FOUND.
C
     420  CALL MP2(A, J1, L2, L20, JL2, IP, ZSCHR, SSCHR, KP, Q1, N, −1)
          IF(IP.NE.0) GO TO 421
          IF(Z.NE.0) GO TO 422
          IF(R.NE.0) GO TO 423
          FALL = 1
          RETURN
     423  KP = R
          L10 = L10 + 1
          L1(L10) = R
          R = 0
          GO TO 420
     422  IP = Z
          GO TO 130
     421  IF(Z.EQ.0) GO TO 130
          KH = Z * ZSCHR + 1
          KH1 = KP * SSCHR + 1
          KH0 = KH + KH1 − 1
          Q3 = − A(KH)/A(KH0)
          IF(Q3.LE.Q1.AND.A(KH1).LE.0.) IP = Z
          GO TO 130
```

```
C
C       PROCEDURE LIST
C       LIST ENTERS ALL ROW INDICES WITH NONNEGATIVE A(I * ZSCHR)
C       INTO LIST L2 AND SETS Z EQUAL TO THAT INDEX OF THE FIRST ROW
C       FOR WHICH A(I * ZSCHR) IS NEGATIVE.
C
  500 L20 = 0
      Z = 0
      DO 501 II = 1, M12
      KH = II * ZSCHR + 1
      IF(A(KH).LT.0.) GO TO 502
      L20 = L20 + 1
      L2(L20) = II
      GO TO 501
  502 IF (Z.EQ.0) Z = II
  501 CONTINUE
      GO TO KW, (102, 301)
      END
```

4.12 ALGOL Program for the Gomory Algorithm

(All independent variables are assumed to be integer valued)

```
procedure    gomory (a, n, m1, m2, m3, zschrt, sschrt, fall, proto1, proto2);
             value n, m1, m2, m3, zschrt, sschrt;
             integer n, m1, m2, m3, zschrt, sschrt, fall;  array a;
             integer array proto1, proto2;
             comment: routine for the minimization of a linear progr⸺ under
             the additional condition that all independent variables are integer
             valued. gomory uses the procedures simplex and dusex. the tableau
             is stored row by row;
             begin
             integer i, k, j, m4;  boolean w;
g01:         for k:=0 step 1 until n do
             a[0 × zschrt + k × sschrt]:= −a[0 × zschrt + k × sschrt];
             simplex (a, zschrt, sschrt, n, m1, m2, m3, fall, proto1, proto2);
             if fall > 0 then go to g04;
             m4:= m1 + m2 + m3;  j:= n;
             for k:= 1 step 1 until n − m3 do
g11:             if proto1[k] > n + m1 + m2 ∧ proto1[k] ≤ n + m4 then
                 begin proto1[k]:= proto1[j];
                 for i:= 0 step 1 until m4 do
                     a[i × zschrt + k × sschrt]:= a[i × zschrt + j × sschrt];
                 j:= j − 1;  go to g11;
                 end k;
             n:= n − m3;
```

```
         for k:= 0 step 1 until n do
         a[0 × zschrt + k × sschrt] := −a[0 × zschrt + k × sschrt];
g02:     for i:= 1 step 1 until m4 do
         if proto2[i] ≤ n∧ (entier(a[i × zschrt]) − a[i × zschrt]) ≠ 0 then
             begin
             m4:= m4 + 1; proto2[m4] := n + m4; w:= true;
             a[m4 × zschrt] := entier(a[i × zschrt]) − a[i × zschrt];
             for k:= 1 step 1 until n do
                 a[m4 × zschrt + k × sschrt] := −a[i × zschrt + k × schrt]−
                 entier (−a[i × zschrt + k × sschrt]);
             go to g03;
             end;
         go to g04;
g03:     dusex (a, zschrt, sschrt, n, m4, fall, w, proto1, proto2);
         if fall > 0 then go to g04;
         go to g02;
g04:     end gomory;
```

4.13   FORTRAN Program for the Gomory Algorithm

```
C    ROUTINE FOR THE MINIMIZATION OF A LINEAR PROGRAM UNDER THE
C    ADDITIONAL CONDITION THAT ALL INDEPENDENT VARIABLES ARE
C    INTEGER-VALUED. GOMORY USES THE SUBROUTINES SIMPLEX AND
C    DUSEX. THE TABLEAU IS STORED ROW BY ROW.
C
     SUBROUTINE GOMORY(A, J1, N, M1, M2, M3, ZSCHR, SSCHR, FALL,
    1       PROTO1, IPROT1, PROTO2, IPROT2, L1, JL1, L2, JL2, L3, JL3)
     DIMENSION A(J1), L1(JL1), L2 (JL2), L3 (JL3)
     INTEGER PROTO1(IPROT1), PROTO2 (IPROT2), FALL, ZSCHR, SSCHR
     LOGICAL W
     N1 = N + 1
     DO 101 K = 1, N1
     KH = (K − 1) * SSCHR + 1
 101 A(KH) = −A(KH)
     CALL SIMPLX (A, J1, ZSCHR, SSCHR, N, M1, M2, M3, FALL, PROTO1,
    1       IPROT1, PROTO2, IPROT2, L1, JL1, L2, JL2, L3, JL3)
     IF(FALL.GT.0)RETURN
     M4 = M1 + M2 + M3
     DO 102 K = 1, N1
     KH = (K − 1) * SSCHR + 1
 102 A(KH) = −A(KH)
C
 200 DO 201 I = 1, M4
     KH = I * ZSCHR + 1
     Q = ABS(AINT(A(KH)) − A(KH))
     IF(Q.LT.0.)A(KH) = AINT(A(KH))
```

```
201  IF (PROTO2(I).LE.N.AND.AINT(A(KH)) — A(KH).NE.0.) GO TO 202
     RETURN
202  M4 = M4 + 1
     PROTO2(M4) = N + M4
     W = .TRUE.
     KH = M4 * ZSCHR + 1
     KH1 = I * ZSCHR + 1
     A(KH) = AINT(A(KH1)) — A(KH1)
     DO 203 K = 1, N
     KH = M4 * ZSCHR + K * SSCHR + 1
     KH1 = I * ZSCHR + K * SSCHR + 1
     Q = AINT(—A(KH1))
     IF(Q.LT.0..AND.Q.NE.(—A(KH1))) Q = Q — 1.
203  A(KH) = — A(KH1) — Q
C
     CALL DUSEX(A, J1, ZSCHR, SSCHR, N, M4, FALL, W, PROTO1, IPROT1,
1         PROTO2, IPROT2, L1, JL1, L2, JL2)
     IF(FALL.GT.0) RETURN
     GO TO 200
     END
```

4.14 ALGOL Program for the Beále Algorithm

procedure	beale (c, a, n, m, zschrt, sschrt, proto1, proto2, fall); **value** n, m, zschrt, sschrt; **integer** n, m, zschrt, sschrt, fall; **array** c, a; **integer array** proto1, proto2; **comment:** routine for determining the minimum of a definite quadratic form subject to linear constraints. beale uses the global procedures mp2, mp3, mp5, mp8, and mp9; **begin** **integer** i, k, ip, kp, v, s, z, m1, t, r; **real** max, q1; **integer array** l1[0:n], l2[0:n + m], list[0:n], ablist[1:n]; **boolean** b1;

program :	**for** k := 1 **step** 1 **until** n **do** l1[k] := proto1[k] := k;
	for i := 1 **step** 1 **until** m **do** l2[i] := i;
	for i := 1 **step** 1 **until** m **do** proto2[i] = n + i;
	l1[0] := n;
	l2[0] := m;
	for k := 1 **step** 1 **until** n **do** ablist[k] := 0;
	list[0] := 0; m1 := m;
	comment: the block beale1 determines the pivotcolumn kp;

beale1 :	**if** list[0] ≠ 0 **then**
	        begin
	        mp8(c, 0, 0, 1, list, kp, max);

```
          if max ≠ 0 then go to beale2;
          end;
          mp5 (c, l1, q1, kp, 0, 0, 1);
          if q1 ≥ 0 then
              begin
              fall := 0;
              go to beale6;
              end;
          comment: if all leading elements of u-columns equal zero and no
          leading element of an x-column is negative, the minimum has been
          obtained;
          comment: the block beale2 determines the pivotrow;
beale2:   max := if c[kp × (n + 1) − (kp × (kp − 1))/2] > 0 then
              c[kp]/c[kp × (n + 1) − (kp × (kp − 1))/2] else 0;
          v := if max > 0 then 1 else −1;
          mp2(a, l2, ip, zschrt, sschrt, kp, q1, n, v);
          if ip = 0 ∧ max = 0 then
              begin
              fall := 1;
              go to beale6;
              end;
          comment: if ip and max are equal to zero, no finite solution
          exists;
          if ip = 0 then go to beale4;
beale3:   if q1 ≤ abs(max) then
              begin
              i := proto1[kp];
              proto1[kp] := proto2[ip];
              proto2[ip] := i;
              mp3 (a, 1, m, 0, n, ip, kp, zschrt, sschrt, 1, 1);
              b1 := true;
              mp9 (a, c, ip, kp, n, m1, zschrt, sschrt, b1);
              if proto2[ip] > n + m1 then
                  begin proto2[ip] := proto2[m]; l2[0] := l2[0] − 1;
                  l1[0] := l1[0] + 1; l1[l1[0]] := kp;
                  t := list[0]; list[0] := t − 1;
                  for k := 1 step 1 until list[0] do if list[k] = kp then t := k;
                  for k := t step 1 until list[0] do list[k] := list[k + 1];
                  for k := 0 step 1 until n do
                      a[ip * zschrt + k * sschrt] := a[m * zschrt + k * sschrt];
                  m := m − 1;
                  end;
              go to beale1;
              end;
          comment: if q1 ≤ max, the pivotrow is contained in the constraint
          section;
beale4:   ablist [kp] := ablist[kp] + 1; t := 0; r := 0;
```

```
beale4a:        if list[0] ≠ 0 then
                for i := 1 step 1 until list[0] do
                    if list[i] = kp then
                        begin
                        z := 0;
                        for s := 0 step 1 until kp do
                            begin
                            a[t × zschrt + s × sschrt] := c[kp + z];
                            z := z + n − s;
                            end s;
                        for s := 1 step 1 until n − kp do
                        a[t × zschrt + (kp + s) × sschrt] := c[kp × (n + 1) −
                        (kp × (kp − 1))/2 + s];
                        ip := t;
                        mp3 (a, r, m, 0, n, ip, kp, zschrt, sschrt, 1, 1);
                        b1 := false;
                        mp9 (a, c, ip, kp, n, m1, zschrt, sschrt, b1);
                        go to beale1;
                        end;
beale5:         list[0] := list[0] + 1;
                list[list[0]] := kp;
                proto2[m + 1] := proto1[kp];
                proto1[kp] := n + m1 + kp;
                l2[m + 1] := m + 1;
                l2[0] := m + 1;
                for s := 1 step 1 until l1[0] do if l1[s] = kp then t := s;
                for s := t step 1 until l1[0] do l1[s] := l1[s + 1];
                l1[0] := l1[0] − 1;
                m := t := m + 1; r := 1;
                go to beale4a;
beale6:         end beale;
```

4.15   FORTRAN Program for the Beale Algorithm

```
C       ROUTINE FOR DETERMINING THE MINIMUM OF A DEFINITE QUADRATIC
C       FORM SUBJECT TO LINEAR CONSTRAINTS. BEALE USES THE SUBROU-
C       TINES MP2, MP3, MP5, MP8, and MP9.
C
        SUBROUTINE BEALE (C, JC, A, J1, N, M, ZSCHR, SSCHR, PROTO1, IPROT1,
       1     PROTO2, IPROT2, LIST, JLIST, ABLIST, JABLIS, L1, JL1, L2, JL2, FALL)
        INTEGER ZSCHR, SSCHR, FALL, PROTO1, PROTO2, ABLIST, V, S, Z, T, R
        DIMENSION C(JC), A(J1), PROTO1(IPROT1), PROTO2(IPROT2), LIST(JLIST),
       1     ABLIST(JABLIS), L1(JL1), L2(JL2)
        REAL MAX
        LOGICAL B1
```

```
        DO 1 K = 1, N
        L1(K) = K
      1 PROTO1(K) = K
        DO 2 I = 1, M
        L2(I) = I
      2 PROTO2(I) = N + I
        L10 = N
        L20 = M
        DO 3 K = 1, N
      3 ABLIST(K) = 0
        LIST0 = 0
        M1 = M
C
C       SECTION 1XXX DETERMINES THE PIVOTCOLUMN KP.
C
   1000 IF(LIST0.EQ.0) GO TO 1001
        CALL MP8 (C, JC, 0, 0, 1, LIST, LIST0, JLIST, KP, MAX)
        IF(MAX.NE.0.) GO TO 2000
   1001 CALL MP5(C, JC, L1, L10, JL1, Q1, KP, 0, 0, 1)
        IF(Q1.LT.0.) GO TO 2000
        FALL = 0
        RETURN
C
C       IF ALL LEADING ELEMENTS OF THE U-COLUMNS EQUAL ZERO AND NO
C       LEADING ELEMENT OF AN X-COLUMN IS NEGATIVE, THE MINIMUM
C       HAS BEEN OBTAINED.
C
C       SECTION 2XXX DETERMINES THE PIVOTROW
C
   2000 KH = KP * (N + 1) − ((KP − 1) * KP)/2 + 1
        MAX = 0.
        IF(C(KH).GT.0.)MAX = C(KP + 1)/C(KH)
        V = −1
        IF(MAX.GT.0.)V = 1
        CALL MP2(A, J1, L2, L20, JL2, IP, ZSCHR, SSCHR, KP, Q1, N, V)
        IF(IP.NE.0.OR.MAX.NE.0.) GO TO 3000
        FALL = 1
        RETURN
C
C       IF IP AND MAX EQUAL ZERO, NO FINITE SOLUTION EXISTS
C
   3000 IF(IP.EQ.0..OR.Q1.GT.ABS(MAX)) GO TO 4000
        I = PROTO1(KP)
        PROTO1(KP) = PROTO2(IP)
        PROTO2(IP) = I
        CALL MP3(A, J1, 1, M, 0, N, IP, KP, ZSCHR, SSCHR, 1, 1)
        B1 = .TRUE.
```

```
       CALL MP9(A, J1, C, JC, IP, KP, N, M1, ZSCHR, SSCHR, B1)
       GO TO 6000
C
C      IF Q1 IS NOT LARGER THAN MAX, THE PIVOTROW IS CONTAINED IN
C      THE CONSTRAINT SECTION
C
  4000 ABLIST(KP) = ABLIST(KP) + 1
       T = 0
       R = 0
  4001 IF(LIST0.EQ.0) GO TO 5000
       DO 4002 I = 1, LIST0
  4002 IF(LIST(I).EQ.KP) GO TO 4003
       GO TO 5000
  4003 Z = 0
       KKP = KP + 1
       DO 4004 S = 1, KKP
       IS = S - 1
       KH = T * ZSCHR + IS * SSCHR + 1
       KH1 = KP + Z + 1
       A(KH) = C(KH1)
  4004 Z = Z + N - IS
       NKP = N - KP
       IF(NKP.LT.1) GO TO 4005
       DO 4006 S = 1, NKP
       KH = T * ZSCHR + (KP + S) * SSCHR + 1
       KH1 = KP * (N + 1) - (KP * (KP - 1))/2 + S + 1
  4006 A(KH) = C(KH1)
  4005 IP = T
       CALL MP3(A, J1, R, M, 0, N, IP, KP, ZSCHR, SSCHR, 1, 1)
       B1 = .FALSE.
       CALL MP9(A, J1, C, JC, IP, KP, N, M1, ZSCHR, SSCHR, B1)
       GO TO 1000
C
  5000 LIST0 = LIST0 + 1
       LIST(LIST0) = KP
       PROTO2(M + 1) = PROTO1(KP)
       PROTO1(KP) = N + M1 + KP
       L2(M + 1) = M + 1
       L20 = M + 1
       DO 5001 S = 1, L10
  5001 IF(L1(S).EQ.KP) T = S
       DO 5003 S = T, L10
  5003 L1(S) = L1(S + 1)
       L10 = L10 - 1
       M = M + 1
       T = M
       R = 1
       GO TO 4001
```

```
6000 IF (PROTO2(IP).LE.N + M1) GO TO 1000
     PROTO2(IP) = PROTO2(M)
     L20 = L20 − 1
     L10 = L10 + 1
     L1(L10) = KP
     LIST0 = LIST0 − 1
     DO 6001 K = 1, LIST0
6001 IF (LIST(K).EQ.KP) GO TO 6003
     GO TO 6004
6003 DO 6005 I = K, LIST0
6005 LIST(I) = LIST(I + 1)
6004 NN = N + 1
     KH1 = IP ∗ ZSCHR + 1
     KH2 = M ∗ ZSCHR + 1
     DO 6002 K = 1, NN
     KH0 = (K − 1) ∗ SSCHR
     KH3 = KH1 + KH0
     KH4 = KH2 + KH0
6002 A(KH3) = A(KH4)
     M = M − 1
     GO TO 1000
     END
```

4.16 ALGOL Program for the Wolfe Algorithm

```
procedure   wolfe (a, x, zschrt, sschrt, n, m, proto1, proto2, fall)
            value zschrt, sschrt, n, m;
            integer zschrt, sschrt, n, m, fall;
            array a, x; integer array, proto1, proto2;
            comment: routine for computing the minimum of a quadratic
            form subject to linear constraints by means of the long form of
            wolfe's method. wolfe uses the global procedures mp2, mp3, and
            mp5;
            begin
            integer i, k, ip, kp, s, t, n1, v;
            integer array l1[0:3 × n + m + 1], l2[0:m + n],
            list1[0:m + n + 1];
            real q1, q2;
        procedure   mp11 (a, zschrt, sschrt, n, m, list1);
            value m, zschrt, sschrt; integer n, m, zschrt, sschrt;
            array a; integer array list1;
            comment: removal of the columns of tableau a contained in list 1;
            begin
            integer i, k, s, t;
            t := 0;
```

```
        for k: = 1 step 1 until list1[0] do
            begin
            for s: = list1[k] step 1 until n − 1 do
            for i: = 0 step 1 until m do
            a[i × zschrt + (s − t) × sschrt]: =
                a[i × zschrt + (s − t + 1) × sschrt];
            t: = t + 1;
            end k;
            n: = n − list1[0];
            end mp11;
    procedure   mp12 (proto1, proto2, l1, m, n, n1);
        value m, n, n1; integer m, n, n1;
        integer array proto1, proto2, l1;
        comment: generation of the list of all columns admissible as pivot-
        columns;
        begin
        integer k;
        l1[0]: = 0;
        for k: = 1 step 1 until n1 do
            begin
            if proto1[k] > n then go to mp121;
            for i: = 1 step 1 until m + n do
            if proto1[k] + n = proto2[i] then go to mp123;
mp122:      l1[0]: = l1[0] + 1;
            l1[l1[0]]: = k;
            go to mp123;
mp121:      if proto1[k] > 2 × n then go to mp122;
            for i: = 1 step 1 until m + n do
            if proto1[k] − n = proto2[i] then go to mp123;
            go to mp122;
mp123:          end k;
            end mp12;
            comment: initialization of the indices and index lists;
program:    for k: = 1 step 1 until 3 × n + m + 1 do
            proto1[k]: = k;
            for i: = 1 step 1 until m + n do
            proto2[i]: = 3 × n + m + 1 + i;
            for k: = 1 step 1 until n, 2 × n + m + 1 step 1 until 3 × n + m do
            if k ≤ n then l1[k]: = k else l1[k − n − m]: = k;
            l1[0]: = 2 × n;
            for i: = 1 step 1 until m + n do
            l2[i]: = i;
            l2[0]: = m + n;
            v: = −1; n1: = 3 × n + m + 1;
            comment: first phase of the wolfe method. part 1a. routine for
            computing the auxiliary objective function;
w1:         for k: = 0 step 1 until n do
```

```
          begin
          q1 := 0;
          for i := 1 step 1 until m do
          q1 := q1 + a[i × zschrt + k × sschrt];
          a[0 × zschrt + k × sschrt] := q1;
          end k;
     for k := n + 1 step 1 until 3 × n + m + 1 do
          a[0 × zschrt + k × sschrt] := 0;
     comment: part 1b. computation of the minimum of the auxiliary
     objective function by the simplex method;
```

w11:
```
     mp5 (a, l1, q1, kp, 0, zschrt, sschrt);
     if q1 ≥ 0 then go to w12;
     comment: if q1 ≥ 0, then the minimum of the auxiliary objective
     function has been obtained;
     mp2 (a, l2, ip, zschrt, sschrt, kp, q1, 3 × n + m + 1, v);
     if ip = 0 then begin fall := 2; go to w4; end;
     i := proto1[kp];
     proto1[kp] := proto2[ip];
     proto2[ip] := i;
     mp3 (a, 0, m + n, 0, 3 × n + m + 1, ip, kp, zschrt, sschrt, 1, 1);
     go to w11;
     comment: part 1c. deletion of the w- and z-columns not contained
     in the basis;
```

w12:
```
     list1[0] := 0; i := 0;
     for k := 1 step 1 until n1 − 1 do
          begin
```

w121:
```
          if k > n1 − 1 − i then go to w122;
          if proto1[k] ≥ 2 × n + m + 1 ∧ proto1[k] ≤ 3 × n + m ∨
          proto1[k] ≥ 3 × n + m + 2 ∧ proto1[k] ≤ 3 × n + 2 × m + 1 ∨
          proto1[k] ≥ 3 × n + 2 × m + 2 ∧ proto1[k] ≤ 4 × n + 2 × m + 1
          then
               begin
               list1[0] := list1[0] + 1;
               list1 [list1[0]] := k + i;
               for s := k step 1 until n1 − 1 − i do
               proto1[s] := proto1[s + 1];
               i := i + 1;
               go to w121;
               end;
          end k;
```

w122:
```
     mp11 (a, zschrt, sschrt, n1, m + n, list1);
     comment: start of the second phase of the wolfe method.
     part 2a. computation of the second auxiliary objective function;
```

w2:
```
     for k := 0 step 1 until n1 do
          begin
          q1 := 0;
          for i := 1 step 1 until n + m do
          if proto2[i] ≥ 2 × n + m + 1 ∧ proto2[i] ≤ 3 × n + m ∨
```

$proto2[i] \geq 3 \times n + 2 \times m + 2 \wedge proto2[i] \leq 4 \times n + 2 \times m + 1$
then $q1 := q1 + a[i \times zschrt + k \times sschrt];$
$a[0 \times zschrt + k \times sschrt] := q1;$
end $k;$
comment: *part 2b. determination of the minimum of the second*
auxiliary objective function;

w21: $mp12\ (proto1,\ proto2,\ l1,\ m,\ n,\ n1 - 1);$
$mp5\ (a,\ l1,\ q1,\ kp,\ 0,\ zschrt,\ sschrt);$
if $q1 \geq 0$ **then go to** w22;
$mp2\ (a,\ l2,\ ip,\ zschrt,\ sschrt,\ kp,\ q1,\ n1,\ v);$
if $ip = 0$ **then begin** $fall := 2;$ **go to** w4; **end;**
$i := proto1[kp];$
$proto1[kp] := proto2[ip];$
$proto2[ip] := i;$
$mp3\ (a,\ 0,\ m + n,\ 0,\ n1,\ ip,\ kp,\ zschrt,\ sschrt,\ 1,\ 1);$
go to w21;
comment: *part 2c. deletion of the z-columns from the tableau;*

w22: $i := list1[0] := 0;$
for $k := 1$ **step** 1 **until** $n1 - 1$ **do**
              begin
w221:   **if** $k > n1 - 1 - i$ **then go to** w222;
              if $proto1[k] \geq 2 \times n + m + 1 \wedge proto1[k] \leq 3 \times n + m \vee$
              $proto1[k] \geq 3 \times n + 2 \times m + 2 \wedge$
              $proto1[k] \leq 4 \times n + 2 \times m + 1$ **then**
                   begin
                   $list1[0] := list1[0] + 1;$
                   $list1[list1[0]] := k + i;$
                   for $s := k$ **step** 1 **until** $n1 - 1 - i$ **do**
                   $proto1[s] := proto1[s + 1];$
                   $i := i + 1;$
                   go to w221;
                   end;
              end $k;$
w222: $mp11\ (a,\ zschrt,\ sschrt,\ n1,\ m + n,\ list1);$
              for $k := 0$ **step** 1 **until** $n1 - 1$ **do** $a[0 \times zschrt + k \times sschrt] := 0;$
              $a[0 \times zschrt + n1 \times sschrt] := -1;$
              comment: *part 3. start of the third phase of the wolfe method;*
w3: $mp12\ (proto1,\ proto2,\ l1,\ m,\ n,\ n1);$
              $x[0] := a[0 \times zschrt + 0 \times sschrt];$
              for $i := 1$ **step** 1 **until** $m + n$ **do** $x[i] := 0;$
              for $i := 1$ **step** 1 **until** $m + n$ **do if** $proto2[i] \leq n$ **then**
                   $x[proto2[i]] := a[i \times zschrt + 0 \times sschrt];$
              $mp5\ (a,\ l1,\ q1,\ kp,\ 0,\ zschrt,\ sschrt);$
              if $q1 \geq 0$ **then begin** $fall := 2;$ **go to** w4; **end;**
              $mp2\ (a,\ l2,\ ip,\ zschrt,\ sschrt,\ kp,\ q1,\ n1,\ v);$
              if $ip = 0$ **then begin** $fall := 2;$ **go to** w4; **end;**
              $i := proto1[kp];$
              $proto1[kp] := proto2[ip];$

```
            proto2[ip] := i;
            mp3 (a, 0, m + n, 0, n1, ip, kp, zschrt, sschrt, 1, 1);
            if −a[0 × zschrt + 0 × sschrt] < 1 then go to w3;
            comment: computation of the solution;
  w33:      q1 := (−a[0] − 1)/(−a[0] + x[0]);
            q2 := (1 + x[0])/(−a[0] + x[0]);
            for i := 1 step 1 until n do x[i] := q1 × x[i];
            for i := 1 step 1 until m + n do
                if proto2[i] ≤ n then
                    x[proto2[i]] := x[proto2[i]] + q2 × a[i × zschrt];
            fall := 0;
  w4:      end wolfe;
```

4.17 FORTRAN Program for the Wolfe Algorithm

```
C       ROUTINE FOR COMPUTING THE MINIMUM OF A QUADRATIC FORM
C       SUBJECT TO LINEAR CONSTRAINTS BY MEANS OF THE LONG FORM OF
C       THE WOLFE METHOD. WOLFE USES THE SUBROUTINES MP2, MP3, AND
C       MP5.
C
        SUBROUTINE WOLFE (A, J1, X, X0, JX, ZSCHR, SSCHR, N, M, PROTO1,
       1          IPROT1, PROTO2, IPROT2, FALL, L1, JL1, L2, JL2, LIST1, JLIST1)
        DIMENSION A(J1), L1(JL1), L2(JL2), X(JX), LIST1(JLIST1)
        INTEGER FALL, PROTO1(IPROT1), PROTO2(IPROT2), S, T, ZSCHR, SSCHR, V
C
C       INITIALIZATION OF THE INDICES AND INDEX LISTS
C
        N1 = 3 * N + M + 1
        DO 1 K = 1, N1
      1 PROTO1(K) = K
        NM = N + M
        DO 2 I = 1, NM
        PROTO2(I) = N1 + I
      2 L2(I) = I
        L20 = NM
        DO 3 K = 1, N
      3 L1(K) = K
        NN = 2 * N + M + 1
        NN1 = N1 − 1
        DO 4 K = NN, NN1
        J = K − NM
      4 L1(J) = K
        L10 = 2 * N
        V = −1
C
C       PHASE 1 OF THE WOLFE METHOD
```

```
C
C       ROUTINE FOR COMPUTING THE AUXILIARY OBJECTIVE FUNCTION
C
 1000 NNN = N + 1
      DO 1001 KK = 1, NNN
      K = KK − 1
      Q1 = 0.
      KH = K ∗ SSCHR + 1
      DO 1002 I = 1, M
      KH0 = KH + I ∗ ZSCHR
 1002 Q1 = Q1 + A(KH0)
 1001 A(KH) = Q1
      DO 1003 K = NN, N1
      KH = K ∗ SSCHR + 1
 1003 A(KH) = 0.
C
C       COMPUTATION OF THE MINIMUM OF THE AUXILIARY OBJECTIVE
C       FUNCTION BY THE SIMPLEX METHOD
C
 1100 CALL MP5(A, J1, L1, L10, JL1, Q1, KP, 0, ZSCHR, SSCHR)
      IF(Q1 . GE . 0.) GO TO 1200
C
C       IF Q1 IS NONNEGATIVE, THE MINIMUM OF THE AUXILIARY OBJECTIVE
C       FUNCTION HAS BEEN OBTAINED.
C
      CALL MP2(A, J1, L2, L20, JL2, IP, ZSCHR, SSCHR, KP, Q1, N1, V)
      IF(IP . NE . 0) GO TO 1101
      FALL = 2
      RETURN
 1101 I = PROTO1(KP)
      PROTO1(KP) = PROTO2(IP)
      PROTO2(IP) = I
      CALL MP3(A, J1, 0, NM, 0, N1, IP, KP, ZSCHR, SSCHR, 1, 1)
      GO TO 1100
C
C       ROUTINE FOR DELETION OF THE W- AND Z-COLUMNS NOT IN THE
C       BASIS.
C
 1200 LIST10 = 0
      I = 0
      DO 1201 K = 1, NN1
 1204 IF(K . GT . NN1 − I) GO TO 1203
      IF(PROTO1(K) . LT . NN . OR . PROTO1(K) . EQ . N1) GO TO 1201
      LIST10 = LIST10 + 1
      LIST1(LIST10) = K + I
      NNN = N1 − 1 − I
      DO 1202 S = K, NNN
 1202 PROTO1(S) = PROTO1(S + 1)
      I = I + 1
```

```
           GO TO 1204
     1201  CONTINUE
     1203  ASSIGN 2000 TO KW
           GO TO 9000
C
C          EXECUTION OF MP11
C
C          START OF THE SECOND PHASE OF THE WOLFE METHOD. COMPUTATION
C          OF THE SECOND AUXILIARY OBJECTIVE FUNCTION.
C
     2000  NNN = N1 + 1
           DO 2001 KK = 1, NNN
           K = KK − 1
           Q1 = 0.
           KH0 = K * SSCHR + 1
           DO 2002 I = 1, NM
           KH = KH0 + I * ZSCHR
     2002  IF(PROTO2(I).GE.2 * N + M + 1.AND.PROTO2(I).LE.3 * N+M.OR.PROTO2(I).
          1    GE.3 * N + 2 * M + 2. AND . PROTO2(I) . LE.4 * N + 2 * M + 1)
          2    Q1 = Q1 + A(KH)
     2001  A(KH0) = Q1
C
C          DETERMINATION OF THE MINIMUM OF THE SECOND AUXILIARY
C          OBJECTIVE FUNCTION
C
     2100  NN1 = N1 − 1
           ASSIGN 2200 TO KW
           GO TO 8000
C
C          EXECUTION OF MP12
C
     2200  CALL MP5(A, J1, L1, L10, JL1, Q1, KP, 0, ZSCHR, SSCHR)
           IF(Q1 . GE . 0.) GO TO 2300
           CALL MP2(A, J1, L2, L20, JL2, IP, ZSCHR, SSCHR, KP, Q1, N1, V)
           IF(IP . NE . 0) GO TO 2201
           FALL = 2
           RETURN
     2201  I = PROTO1(KP)
           PROTO1(KP) = PROTO2(IP)
           PROTO2(IP) = I
           CALL MP3(A, J1, 0, NM, 0, N1, IP, KP, ZSCHR, SSCHR, 1, 1)
           GO TO 2100
C
C          DELETION OF THE Z-COLUMNS FROM THE TABLEAU
C
     2300  I = 0
           LIST10 = 0
           NN1 = N1 − 1
           DO 2301 K = 1, NN1
```

```
2306 IF(K.GT.NN1 — I) GO TO 2305
     IF(.NOT.(PROTO1(K).GE.2 * N + M + 1.AND.PROTO1(K).LE.3 * N + M.OR.
   1       PROTO1(K).GE.3 * N + 2 * M + 2)) GO TO 2301
     LIST10 = LIST10 + 1
     LIST1(LIST10) = K + I
     NNN = N1 — I
     DO 2302 S = K, NNN
2302 PROTO1(S) = PROTO1(S + 1)
     I = I + 1
     GO TO 2306
2301 CONTINUE
2305 ASSIGN 2303 TO KW
     GO TO 9000
C
C    EXECUTION OF MP11
C
2303 DO 2304 KK = 1, N1
     K = KK — 1
     KH = K * SSCHR + 1
2304 A(KH) = 0.
     KH = KH + SSCHR
     A(KH) = —1.
C
C    START OF THE THIRD PHASE OF THE WOLFE METHOD.
C
3000 NN1 = N1
     ASSIGN 3200 TO KW
     GO TO 8000
C
C    EXECUTION OF MP12
C
3200 X0 = A(1)
     DO 3204 I = 1, NM
3204 X(I) = 0.
     DO 3201 I = 1, NM
     KH = I * ZSCHR + 1
     KH0 = PROTO2(I)
3201 IF(KH0.LE.N)X(KH0) = A(KH)
     CALL MP5(A, J1, L1, L10, JL1, Q1, KP, 0, ZSCHR, SSCHR)
     IF(Q1.LT.0.) GO TO 3202
3203 FALL = 2
     RETURN
3202 CALL MP2(A, J1, L2, L20, JL2, IP, ZSCHR, SSCHR, KP, Q1, N1, V)
     IF(IP.EQ.0) GO TO 3203
     I = PROTO1(KP)
     PROTO1(KP) = PROTO2(IP)
     PROTO2(IP) = I
     CALL MP3(A, J1, 0, NM, 0, N1, IP, KP, ZSCHR, SSCHR, 1, 1)
     IF(—A(1).LT.1.) GO TO 3000
C
```

```
C       COMPUTATION OF THE SOLUTION
C
  3300  Q2 = A(1) − X0
        Q1 = (A(1) + 1.)/Q2
        Q2 = −(1. + X0)/Q2
        DO 3301 I = 1, N
  3301  X(I) = Q1 * X(I)
        DO 3302 I = 1, NM
        KH = PROTO2(I)
        KH0 = I * ZSCHR + 1
  3302  IF(KH.LE.N)X(KH) = X(KH) + Q2 * A(KH0)
        FALL = 0
        RETURN
C
C       PROCEDURE MP11 ROUTINE FOR REMOVAL OF THE COLUMNS OF
C       TABLEAU A CONTAINED IN LIST 1
C
  9000  IT = 0
        MM = NM + 1
        KE = N1 − 1
        IF(LIST10.LT.1) GO TO 9001
        DO 9002 K = 1, LIST10
        KA = LIST1(K)
        DO 9003 S = KA, KE
        DO 9003 I = 1, MM
        KH = (I − 1) * ZSCHR + (S − IT) * SSCHR + 1
        KH0 = KH + SSCHR
  9003  A(KH) = A(KH0)
  9002  IT = IT + 1
  9001  N1 = N1 − LIST10
        GO TO KW, (2000, 2303)
C
C       PROCEDURE MP12
C       GENERATION OF LIST L1 OF ALL COLUMNS ADMISSIBLE AS PIVOTCOLUMNS
C
  8000  L10 = 0
        DO 8001 K = 1, NN1
        IF(PROTO1(K).GT.N) GO TO 8002
        DO 8003 I = 1, NM
  8003  IF(PROTO1(K) + N.EQ.PROTO2(I)) GO TO 8001
  8005  L10 = L10 + 1
        L1(L10) = K
        GO TO 8001
  8002  IF(PROTO1(K).GT.2 * N) GO TO 8005
        DO 8004 I = 1, NM
  8004  IF(PROTO1(K) − N.EQ.PROTO2(I)) GO TO 8001
        GO TO 8005
  8001  CONTINUE
        GO TO KW, (2200, 3200)
        END
```

List of Existing Computer Codes

Computer	Problem	Name of the routine[2]
Bull-General Electric		
GAMMA 30	Lin. Pr.	PFI
GE 400	Lin. Pr.	Linear programming package
GE 600	Lin. Pr.	LP/600
Control Data		
1604/3400/3600/3800	Lin. Pr.	CDM 4 (Control Data Math. Progr. System 4)
1604/all 3000 and 6000 models	Integer Pr.	ILP 1 (Integer linear programming 1)
3100/3200/3300/3500	Lin. Pr.	REGINA I
3600/3800	Lin. Pr.	OPHELIE
6400/6500/6600	Lin. Pr.	OPTIMA
IBM		
7040/7044	Lin. Pr.	7040/44 LINEAR PROGRAMMING SYSTEM
7090/7094	Lin. Pr.	7090/94 LINEAR PROGRAMMING SYSTEM
7094	Lin. Pr. extended	LINEAR AND SEPARABLE PROGRAMMING SYSTEM
7044	Quadr. Pr.	QUADRATIC PROGRAMMING CODE
7040/7090	Nonlin. Pr.	RAC NONLINEAR PROGRAM SEQUENTIAL UNCONSTRAINED MINIMIZATION TECHNIQUE

[1] Many computer manufacturers provide ready library codes for problems of mathematical optimization.

[2] German routines translated into English.

on Mathematical Programming[1]

Language	Method	Maximal number of	
		Constraints	Variables
Autocoder	Rev. Sx., prod. form	500	1000
FORTRAN	Rev. Sx., prod. form	750	1200
FORTRAN	Rev. Sx., prod. form	4096	Unlimited
FORTRAN	Rev. Sx. with slack var.	400	900
FORTRAN	Gomory, Tucker-transf.	Core memory limitation	
FORTRAN/Ass.	Rev. Sx., prod. form	1024	Unlimited
FORTRAN/Ass.	Rev. Sx., prod. form	4095	Unlimited
Assembler	Rev. Sx., decomposition	4095	Unlimited
MAP	Rev. Sx., extended	1023	200,000
FAP	Rev. Sx., extended	1023	200,000
FORTRAN	Rev. Sx., extended	300	4000
FORTRAN IV		(Const., + var. = 508)	
FORTRAN IV	Cf. Manag. Sc. 10/2 (64)		

(continued)

Computer	Problem	Name of the routine[2]
IBM (contd.)		
7090/7094	Nonlin. Pr.	NONLINEAR PROGRAMMING SUBJECT TO LINEAR CONSTRAINTS
7090/7094	Integer Pr.	INTEGER LINEAR PROGRAMMING ONE
7090/7094	Transport.	TRANSPORTATION PROBLEM SUBROUTINE
1620	Lin. Pr.	LINEAR PROGRAMMING SYSTEM
1620	Transport.	TRANSPORTATION PROBLEM
7070	Lin. Pr.	LINEAR PROGRAMMING CODE S 2
7070	Transport.	TRANSPORTATION PROBLEM/ DENNIS TECHNIQUE
1401	Lin. Pr.	LINEAR PROGRAMMING SYSTEM
1401	Lin. Pr.	LINEAR PROGRAMMING, REVISED SIMPLEX
1410	Lin. Pr.	BASIC LINEAR PROGRAMMING
1130	Lin. Pr.	LP-MOSS (Lin. Progr. Math. Opt. Subroutine Syst.)
360	Lin. Pr.	MPS/360 (Math. Programming System)
360	Transport.	TRANSPORTATION PROBLEM
ICT		
1900	Lin. Pr.	XDL0, XDL1, XDL2
1900	Lin. Pr.	XDL4, XDL8
1900	Transport.	XDT2, XDT3
Honeywell		
H200	Lin. Pr.	LINEAR PROGRAMMING D

Language	Method	Maximal number of	
		Constraints	Variables
FORTRAN II	Iterative, derivative of nonlin. fcts. has to be given		
FORTRAN II	Gomory	120	300
		(start: 70/250)	
FORTRAN II	Primal Balinsky-Gomory	Elements in matrix = 10,000	
	Rev. Sx.	100–400	1000–3000
	Stepping stone	Elements in matrix = 40 × (80 to 300)	
	Rev. Sx.	180	800
Autocoder	Stepping stone	Elements in matrix = 50–275 × 500–275	
Autocoder	Simplex, modified	47–97	900
	Rev. Sx.	108	Unlimited
Autocoder	Simplex	150	
FORTRAN/Ass.	Rev. Sx., prod. form	700	Unlimited
Assembler	Rev. Sx., prod. form	4095	Unlimited
FORTRAN	MODI (stepping stone)	Core memory limitation	
—	Rev. Sx.	A and B matrices in core	
—	Rev. Sx.	Matrix on tape	
—	Ford–Fulkerson	Core memory limitation	
FORTRAN	Simplex	48	98

(*continued*)

Computer	Problem	Name of the routine[2]
Honeywell (Contd.)		
H200	Lin. Pr.	LINEAR PROGRAMMING K (and H)
M8200	Lin. Pr.	ALPS
NCR		
315 with 1 CRAM	Lin. Pr.	LP
315	Transport	TRANSPORT
Siemens		
2002	Lin. Pr.	Simplex method for linear programming
4004/35-55	Lin. Pr.	Linear programming—System I
4004/35-55	Lin. Pr.	Linear programming—System II
4004/35-55	Nonlin. Pr.	ALMA
4004/35-55	Quadr. Pr.	Convex programming
4004/35-55	0/1—Pr.	Integer linear optimization with 0/1 limitation
4004/35-55	Integer Pr.	Integer linear optimization
4004/15-55	Transport	TRAP
Sperry Rand–Univac		
1107/1108	Lin. Pr.	1107 LP (compatible with IBM LP 90)
1107/1108	Lin. Pr.	ILONA (4 to 5 times faster than 1107 LP)
1107/1108	Transport.	PFTC
Telefunken		
TR4	Lin. Pr.	KPDA

Language	Method	Maximal number of	
		Constraints	Variables
FORTRAN	Revised simplex, extended	750(500)	1500(1000)
	Rev. Sx., extended	Memory ltd.	Unlimited
	Simplex	300	1500
Assembler	Simplex		
FORTRAN IV	Rev. Sx., extended	4000	Unlimited
FORTRAN IV	Rev. Sx., extended decomp.	Unlimited	Unlimited
FORTRAN IV	Method of gradients	999	9999
	Wolfe	500	2000
FORTRAN IV	E. Balas	Unlimited	Unlimited
FORTRAN IV	Gomory	Core memory limitation	
Assembler	Stepping stone	(totally 1200 nodes)	
FORTRAN IV	Rev. Sx., prod. form	(constr. + var. = 8000)	
FORTRAN IV	Rev. Sx., prod. form extended	(constr. + var. = 8000)	
FORTRAN IV	Primal flow	(Elements in matrix = 400×400)	
ALGOL	Primal-dual-combination	(Constr. $\times$ var. = 12000)	

(continued)

Computer	Problem	Name of the routine[2]
Telefunken (contd.) TR4	Lin. Pr.	TELINOP
Zuse Z23	Integer Pr.	Integer optimization
Z23	Allocation	Approximation to the Allocation Problem
Z23	Lin. Pr.	(several older routines available)
Z25	Lin. Pr.	Linear optimization

For further programs we refer to the following periodicals: Communications of the ACM, (complete ALGOL programs); Unternehmensforschung, (references to manufacturers' programs).

Language	Method	Maximal number of	
		Constraints	Variables
ALGOL	Rev. Sx., upper bounding	(constr. + var. = 5000)	
Freiburger code ALGOL	E. Balas	150	200
ALGOL	Simplex	70	100

BIBLIOGRAPHY

Besides those articles cited in the text, the following list contains additional relevant publications concerning the theory and applications of linear and nonlinear optimization.

[1] Antosiewicz, H. A. (ed.): Proceedings of the second symposium in linear programming. 2 vol. Washington 1955.
[2] Arrow, K. J.; Hurwicz, L.; Uzawa, H.: Studies in linear and nonlinear programming. Stanford, Cal. 1958.
[3] Balinski, M. L.; Gomory, R. E.: A primal method for the assignment and transportation problems. Managem. Sci. 10 (1964) 578–593.
[4] Beale, E. M. L.: An alternative method for linear programming. Proc. Cambridge Philos. Soc. 50 (1954) 513–523.
[5] Beale, E. M. L.: Cycling in the dual simplex algorithm. Nav. Res. Log. Quart. (1955) 269–275.
[6] Beale, E. M. L.: On minimizing a convex function subject to linear inequalities. J. Roy. Stat. Soc. 17 B (1955) 173–184.
[7] Beale, E. M. L.: On quadratic programming. Nav. Res. Log. Quart. 6 (1959) 227–243.
[8] Beale, E. M. L.; Hughes, P. A. B.; Small, R. E. : Experiences in using a decomposition program. Computer 8 (1965) 13–18.
[9] Beckmann, M.: Lineare Planungsrechnung. Ludwigshafen am Rhein 1959 = Fachbücher für Wirtschaftstheorie und Oekonometrie.
[10] Bellmann, R.: Dynamic programming. Princeton, N. J. 1957.
[11] Bereanu, B.: On stochastic linear programming. Rev. math. pures et appl. VIII (1963) 683–697.
[12] Boot, J. C. G.: Quadratic programming (Algorithms – Anomalies-Applications). Amsterdam 1964.
[13] Berge, C.: Théorie des graphes et ses applications. Paris 1958.
[14] Bonnesen, T.; Fenchel, W.: Theorie der konvexen Körper. 1948.
[15] Candler, W.; Townsley, R. J.: The maximization of a quadratic function of variables subject to linear inequalities. Managem. Sci. 10 (1964) 515–523.
[16] Charnes, A.: Optimality and degeneracy in linear programming. Econometrica 20 (1952) 160–170.
[17] Charnes, A.; Cooper, W. W.: The stepping stone method of explaining linear programming calculations in transportation problems. Managem. Sci. 1 (1954/55) 49–69.
[18] Charnes, A.; Cooper, W.W.: Management models and industrial applications of linear programming. New York 1961.
[19] Charnes, A.; Cooper, W. W.: Deterministic equivalents for optimizing and satisficing under chance constraints. Operations Res. 11 (1963) 18–39.
[20] Charnes, A.; Cooper, W. W.; Henderson, A.: An introduction to linear programming. New York 1953.
[21] Charnes, A.; Cooper, W. W.; Kortanek, K.: Duality in semi-infinite programs and some works of Haar and Caratheodory. Managem. Sci. 9 (1963) 209–228.
[22] Dantzig, G. B.: Programming in a linear structure. Washington, D. C. 1948.
[23] Dantzig, G. B.: A proof of the equivalence of the programming problem and the game problem. In: [85] chap. 20, 330–335.

162

[24] Dantzig, G. B.: Maximization of a linear function of variables subject to linear inequalities. In: [85] chap. 21, 339–347.

[25] Dantzig, G. B.: Notes on linear programming. Part VII: The dual simplex algorithm. The RAND Corporation RM–1270 (1954).

[26] Dantzig, G. B.: Block triangular systems in linear programming. The RAND Corporation RM–1273 (1954).

[27] Dantzig, G. B.: Notes on linear programming. Parts VIII, IX, X: Upper bounds, secondary constraints and block trinagularity in linear programming. The RAND Corporation RM–1367 (1954); Econometrica 23 (1955) 174–183.

[28] Dantzig, G. B.: Linear programming under uncertainty. Managem. Sci. 1 (1955) 197–206.

[29] Dantzig, G. B.: Recent advances in linear programming. Managem. Sci. 2 (1955/56) 131–144.

[30] Dantzig, G. B.: Discrete variable extremum problems. Operations Res. 5 (1957) 266–277.

[31] Dantzig, G. B.: General convexe objective forms. In: K. J. Arrow, s. Karlin and P. Suppes (eds.), Mathematical methods in the social sciences. Stanford, Cal. 1960. 151–158.

[32] Dantzig, G. B.: On the significance of solving linear programming problems with some integer variables. Econometrica 28 (1960) 30–44.

[33] Dantzig, G. B.: Lineare Programmierung und Erweiterungen. Berlin – Heidelberg – New York 1966.

[34] Dantzig, G. B.; Ford, L. R.; Fulkerson, D. R.: A primal-dual algorithm for linear programs. The RAND Corporation P–778 (1955); und in: H. W. Kuhn and A. W. Tucker (eds.), Linear inequalities and related systems. Princeton, N. J. 1956. 171–181. = Annals of Mathematics Study. No. 38.

[35] Dantzig, G. B.; Fulkerson, D. R.; Johnson, S.: Solution of a large-scale traveling-salesman problem. J. Operations Res. Soc. America 2 (1954) 393–410.

[36] Dantzig, G. B.; Madansky, A.: On the solution of two-stage linear programs under uncertainty. The RAND Corporation P–2039 (1960).

[37] Dantzig, G. B.; Orchard-Hays, W.: Alternate algorithm for the revised simplex method. The RAND Corporation RM–1268 (1953).

[38] Dantzig, G. B.; Orden, A.: Duality theorems. The Rand Corporation RM–1265 (1953).

[39] Dantzig, G. B.; Orden, A.; Wolfe, Ph.: The generalized simplex method for minimizing a linear form under linear inequality restraints. The RAND Corporation RM–1264 (1954); Pacific J. Math. 5 (1955).

[40] Dantzig, G. B.; Wolfe, Ph.: Decomposition principle for linear programs. Operations Res. 8 (1960) 101–111.

[41] Dieter, U.: Programmierung in linearen Räumen. Wahrsch. Rechn. u. verw. Gebiete (1965).

[42] Dorfman, R.: Application of linear programming to the theory of the firm, including an analysis of monopolistic firms by nonlinear programming. Berkeley, Cal. 1951.

[43] Dorfman, R.; Samuelson, P. A.; Solow, R.: Linear programming and economic analysis. New York-Toronto-London 1958.

[44] Dorn, W. S.: Duality in quadratic programming. Quart. Appl. Math. 18 (1960).

[45] Dorn, W. S.: Self-dual quadratic programs. SIAM 9 (1961) 51–54.

[46] Dorn, W. S.: Non-linear programming – a survey. Managem. Sci. 9 (1963) 171–208.

[47] Dresher, M.: Games of strategy: theory and applications. Englewood Cliffs, N. J. 1961.

[48] Dresher, M.: Strategische Spiele. Theorie und Praxis (German ed. by H. P. Künzi) Zürich 1961.

[49] Dresher, M.; Tucker, A. W.; Wolfe, Ph.: Contributions to the theory of games. Vol. III. Princeton, N. J. 1957. = Annals of Mathematics Study. No. 39.

[50] Duane Pyle, L.: The generalized inverse-eigenvector method for solving linear programming problems. Summary. 1964.

[51] Fan, Ky.: Existence theorems and extreme solutions for inequalities concerning convex functions for linear transformations. Math. Z. 68 (1957).

[52] Fenchel, W.: Convex cones, sets, and functions. Lecture notes. Princeton 1953.

[53] Ferguson, A. R.; Dantzig, G. B.: The problem of routing aircraft – a mathematical solution. The RAND Corporation RM–1369 (1954); Aeron. Eng. Rev. 14 (1955) 51–55.

[54] Fiacco, A. V.; McCormick, G. P.: Computational algorithm for the sequential unconstrained minimization technique for nonlinear programming. Managem. Sci. 10 (1964) 601–617.

[55] Fiacco, A. V.; McCormick, G. P.: The sequential unconstrained minimization technique for nonlinear programming, a primal dual method. Managem. Sci. 10 (1964) 360–366.

[56] Fiacco, A. V.; Smith, N. M.; Blackwell, D.: A more general method for nonlinear programming. 1960.

[57] Frank, M.; Wolfe, Ph.: An algorithm for quadratic programming. Nav. Res. Log. Quart. 3 (1956).

[58] Frisch, R.: Principles of linear programming, with particular reference to the double gradient form of the logarithmic potential method. Oslo 1954.

[59] Frisch, R.: The multiplex method for linear and quadratic programming. Oslo 1957.

[60] Fulkerson, D. R.; Dantzig, G. B.: Computation of maximal flows in networks. Nav. Res. Log. Quart. (1955) 277–283.

[61] Gale, D.; Kuhn, H. W.; Tucker, W.: Linear programming and the theory of games. In: [85] chap. 19, 317–329.

[62] Gass, S. I.: Linear programming, methods and applications. New York–Toronto–London 1958.

[63] Goldman, A. J.; Tucker, A. W.: Theory of linear programming. In linear inequalities and related systems. Princeton, N.J. 1956.

[64] Gomory, R. E.: An algorithm for integer solutions to linear programs. Princeton-IBM Math. Res. Project. Techn. Rep. No. 1 (1958).

[65] Gomory, R. E.: Essentials of an algorithm for integer solutions to linear programs. Bull. Amer. Math. Soc. 64 (1958).

[66] Gomory, R. E.: An algorithm for the mixed integer problem. The RAND Corporation P–1885 (1960): and Abstract 553–190: Extention of an Algorithm for Integer-Solutions to Linear Programs. Amer. Math. Soc. Not. 6, No. 1, Issue 36 (1959).

[67] Gomory, R. E.; Hu, T. C.: Multi-terminal network flows. IBM Res. Rep. RC–318 (1960).

[68] Gomory, R. E.; Hu, T. C.: An application of generalized linear programming to network flows. IBM Res. Rep. 1960.

[69] Graves, R. L.; Wolfe, Ph. (eds): Recent advances in mathematical programming. New York–Toronto–London 1963.

[70] Hadley, G.: Linear algebra. Reading, Mass. – Palo Alto–London 1961.

[71] Hadley, G.: Linear programming. Reading, Mass. – Palo Alto – London 1961.

[72] Hadley, G.: Nonlinear and dynamic programming. Reading, Mass. – Palo Alto – London 1964.

[73] Harrison, J. O.: Linear programming and operations research. In: J. F. McCloskey and F. N. Trefethen (eds.), Operations research for management. vol. I. Baltimore 1954. 217–237.

[74] Hartley, H. O.; Hocking, R. R.: Convex programming by tangential approximation. Managem. Sci. 9 (1963) 600–612.

[75] Hestenes, M. R.; Stiefel, E.: Methods of conjugate gradients for solving linear systems. J. Res. Nat. Bur. Stand. 49 (1952).

[76] Hildreth, C. G.: A quadratic programming procedure. Nav. Res. Log. Quart. 4 (1957).
[77] Houthakker, H. S.: The capacity method of quadratic programming. Econometrica 28 (1960) 62–87.
[78] Hurwicz, L.: Programming in linear spaces. In: [2] 38–102.
[79] Jacobs, W. W.: Military applications of linear programming. In: [1] 1–27.
[80] Jaeger, A.: Introduction to analytic geometry and linear algebra. New York 1961.
[81] Joseph, J. A.: The application of linear programming to weapon selection and target analysis. Operations Analysis Technical Memorandum. Washington, D. C. 1954.
[82] Karlin, S.: Mathematical methods and theory in games, programming, and economics. Vol. I and II. Reading. Mass. – Palo Alto – London 1959.
[83] Katzman, I.: Solving feed problems through linear programming. J. Farm. Econ. 38 (1956) 420–429.
[84] Kelley Jr., J. E.: The cutting plane method for solving convex programs. SIAM 8 (1960) 703–712.
[85] Koopmans, T. C. (ed.): Activity analysis of production and allocation. New York 1951. = Cowles Commission Monograph. No. 13.
[86] Krelle, W.: Ganzzahlige Programmierungen; Theorie und Anwendungen in der Praxis. Unternehmensforschung 2 (1958) 161–175.
[87] Krelle, W.; Künzi, H. P.: Lineare Programmierung. Zürich 1959.
[88] Kretschmer, K. S.: On the scope of the theory of programs in paired spaces. International Symposium on Mathematical Programming London 1964.
[89] Kuhn, H. W.; Tucker, A. W.: Non-linear programming. In J. Neyman (ed.): Proceedings of the Second Berkeley Symposium on Math. Stat. and Probab. Berkeley, Cal. 1950. 481–492; abstrakt: Econometrica, 19 (1951) 50–51.
[90] Kuhn, H. W.; Tucker, A. W.: Contributions of theory of games. Vol. 1. Princeton, N. J. 1950. = Annals of Mathematics Study. No. 24.
[91] Künzi, H. P.: Die Simplexmethode zur Bestimmung einer Ausgangslösung bei bestimmten linearen Programmen. Unternehmensforschung 2 (1958) 60–69.
[92] Künzi, H. P.: Nichtlineare Programmierung. ZAMM 41 (1961).
[93] Künzi, H. P.: Die Duoplex-Methode. Unternehmensforschung 7 (1963) 103–116.
[94] Künzi, H. P.: The duoplex method in nonlinear programming, SIAM (1966).
[95] Künzi, H. P.; Krelle, W.: Nichtlineare Programmierung. Berlin–Heidelberg–New York 1962.
[96] Künzi, H. P.; Oettli, W.: Une méthode de résolution de programmes quadratiques en nombres entiers. C. r. Acad. Scie. 252 (1961).
[97] Künzi, H. P.; Tan, S.: Lineare Optimierung großer Systeme. Berlin–Heidelberg–New York 1966. = Lecture notes in mathematics.
[98] Künzi, H. P.; Tzschach, H. G.: The duoplex-algorithm. Numer. Math. 7 (1965).
[99] Lambert, F.: Programmes linéaires mixtes. Cahiers centre d'études rech. oper. (1960).
[100] Land, A. H.; Doig, A. G.: An automatic method of solving discrete programming problems (unpublished work) London 1957.
[101] Lemke, C. E.: The dual method of solving the linear programming problem. Nav. Res. Log. Quart. 1 (1954) 36–47.
[102] Madansky, A.: Some results and problems in stochastic linear programming. The RAND Corporation P-1596 (1959).
[103] Madansky, A.: Inequalities for stochastic linear programming problems. Managem. Sci. 6, No. 2 (1960) 197–204.
[104] Madansky, A.: Methods of solution of linear programs under uncertainty. Operations Res. (1962) 463–470.
[105] Mangasarian, O. L.: Nonlinear programming, problems with stochastic objective functions. Managem. Sci. 10 (1964) 353–359.

[106] Manne, A. S.: Notes on parametric linear programming. The RAND Corporation P–468 (1953).

[107] Massé, P.; Gibrat, R.: Applications of linear programming to investments in the electric power industry. Managem. Sci. **3**. No. 1 (1957) 149–166.

[108] Motzkin, T. S.: Beiträge zur Theorie der linearen Ungleichungen. Diss. Zürich. Basel 1933.

[109] Motzkin, T. S.; Raiffa, H.; Thompson, G. L.; Thrall, R. M.: The double description method. In: H. W. Kuhn and A. W. Tucker (eds.), Contributions to the Theory of Games. Vol. II. Princeton, N. J. 1953. = Annals of Mathematics Study No. 28.

[110] Nef, W.: Die Auflösung linearer Programme ohne Kenntnis einer zulässigen Lösung. Unternehmensforschung 8 (1964) 136–141.

[111] Neumann, J. von; Morgenstern, O.: Theory of games and economic behavior. Princeton, N. J. 1947.

[112] Neumann, J. von; Morgenstern, O.: Spieltheorie und wirtschaftliches Verhalten. Würzburg 1961.

[113] Orchard-Hays, W.; Cutler, L.; Judd, H.: Manual for the RAND IBM Code for linear programming on the 704. The RAND Corporation P–842 (1956) 24–26.

[114] Orchard-Hays, W.: The RAND code for the simplex method. The RAND Corporation RM–1269 (1954).

[115] Orchard-Hays, W.: A composite simplex algorithm. – II. The RAND Corporation RM–1275 (1954).

[116] Orchard-Hays, W.: Background, development and extensions of the revised simplex method. The RAND Corporation RM–1433 (1954).

[117] Orchard-Hays, W.: RAND Code for the simplex method. The RAND Corporation RM–1440 (1955).

[118] Orchard-Hays, W.: Evolution of computer codes for linear programming. The RAND Corporation P–810 (1956) 22–24.

[119] Orden, A.: Goldstein, L.: Symposium on linear inequalities and programming. Washington, D. C. 1952.

[120] Prager, W.: On the caterer problem. Managem. Sci. **3**. No. 1 (1956) 15–23.

[121] Reisch, E.; Eisgruber, L.: Bibliography of linear programming and its application to agricultural economic problems. 1960.

[122] Riley, V.; Gass, S. I.: Linear programming and associated techniques. A comprehensive bibliography on linear, nonlinear and dynamic programming. Baltimore, Md. 1958.

[123] Rosen, J. B.: The gradient projection method for nonlinear programming. Part I: Linear constraints. SIAM 8 (1960) 181–217.

[124] Rosen, J. B.: The gradient projection method for nonlinear programming. Part II: Nonlinear constraints. The RAND Symposium on Math. Prog. Proc. of a Conf. R–351 (1960); SIAM **9** (1961) 514–532.

[125] Rosen, J. B.; Ornea, J.C.: Solution of nonlinear programming problems by partitioning. Management Sci. **10** (1964) 164–173.

[126] Rosen, J. B.; Suzuki, S.: Construction of nonlinear programming test problems. Comm. ACM. 8, 2. 113.

[127] Saaty, T. L.: The number of vertices of a polyhedron. Amer. Math. Month. **65**, No. 5 (1955) 327–331.

[128] Sasieni, M. W.; Yaspan, A.; Friedman, L.: Operations Research: Methods and Problems. New York 1959.

[129] Sasieni, M.; Yaspan, A.; Friedman, L.: Methoden und Probleme der Unternehmensforschung (Transl. by H. P. Künzi). Würzburg 1962.

[130] Shetty, C. M.: A simplified procedure for quadratic programming. Operations Res. **11** (1963) 248–260.

[131] Stiefel, E.: Note on Jordan elimination, linear programming and Tchebycheff approximation. Numer. Math. 2 und 4 (1960).

[132] Stiefel, E.: Relaxationsmethoden bester Strategie zur Lösung linearer Gleichungssysteme. Comment. Math. Helv. **29** (1955).
[133] Stiefel, E.: Einführung in die numerische Mathematik. 3. Aufl. Stuttgart 1965.
[134] Stigler, G. J.: The cost of subsistence. J. Farm. Econ. **27** (1945) 303–314.
[135] Symonds, G. H.: Linear programming. The solution of refinery problems. Esso Standard Oil Company 1955.
[136] Theil, H.; Van de Panne, C.: Quadratic programming as an extension of conventional quadratic maximization. Managem. Sci. **7** (1961) 1–20.
[137] Thrall, R.: Some results in non-linear programming. Part. I. The RAND Corporation RM–909 (1952); [1] Vol. 2. 471–493.
[138] Thrall, R.: Some results in non-linear programming. Part II. The RAND Corporation RM–935 (1952).
[139] Samuelson, P. A.: Linear programming and economic theory. In: [1] vol. 2. 251–272.
[140] Tintner, G.: Stochastic linear programming with applications to agricultural economics. In: [1] Vol. 1. 197–228.
[141] Tucker, A. W.: Dual systems of homogeneous linear relations. In: H. W. Kuhn, A. W. Tucker: Linear inequalities and related systems. Princeton, N. J. 1956. 3–18. = Annals of Mathematics Study. No. 38.
[142] Tucker, A. W.: Linear and non-linear programming. Operations Res. **5** (1957) 244–257.
[143] Tucker, A. W.: Solving a matrix game by linear programming. IBM J. Res. Develop. **4**, No. 5 (1960) 507–517.
[144] Tucker, A. W.: A combinatorial equivalence of matrices. In: R. Bellman and Marshall Hall, Jr. (eds.). Proceedings of Symposia in Applied Mathematics. Vol. X: Combinatorial Analysis. Providence, R. I. 1960. 129–140.
[145] Vajda, S.: The theory of games and linear programming. New York 1956.
[146] Vajda, S.: Theorie der Spiele und Linearprogrammierung. Berlin 1962.
[147] Vajda, S.: Readings in linear programming. London 1958.
[148] Vajda, S.: Lineare Programmierung. Beispiele (German transl. by H. P. Künzi) Zürich 1960.
[149] Vajda, S.: Mathematical programming. Reading, Mass.–Palo Alto–London 1961.
[150] Valentine, F. A.: Convex sets. New York 1964.
[151] Van de Panne, C.: A quadratic programming method. The Mathematical Programming Symposium Chicago 1962.
[152] Vazsonyi, A.: Scientific programming in business and industry. New York 1958.
[153] Wagner, H. M.: A comparison of the original and revised simplex methods. Operations Res. **5** (1957) 361–369.
[154] Wedekind, H.: Primal- und Dual-Algorithmen zur Optimierung von Markov-Prozessen. Unternehmensforschung **8** (1964) 128–135.
[155] Wets, R.: Programming under uncertainty the complete problem. Boeing Scientific Research Laboratories. Mathematic Research. Oct. 1964.
[156] Witzgall, Ch.: Gradient-projection methods for linear programming. IBM Math. Res. Proj. Techn. Rep. **2** (1960).
[157] Witzgall, Ch.: On the gradient projection methods of R. Frisch and J. B. Rosen. The Mathematical Programming Symposium Chicago 1962.
[158] Wolfe, Ph.: The simplex method for quadratic programming. Econometrica **27** (1959) 382–398.
[159] Wolfe, Ph.: A duality theorem for non-linear programming. The RAND Corporation P–2028 (1960).
[160] Wolfe, Ph.: Recent developments in non-linear programming. Part I. The RAND Corporation P–2063 (1960).
[161] Wolfe, Ph.: Some simplex-like non-linear programming procedure. In: Proceedings of the Second Intern. Conf. on O. R. Aix-en-Provence 1960; Operations Res. **10** (1962) 438–447.

[162] Wolfe, Ph.: Methods of nonlinear programming. Mathematical programming Symposium Chicago 1962.
[163] Zoutendijk, G.: Studies in non-linear programming. Some remarks about the gradient projection method of non-linear programming. Koninklije/Shell-Laboratorium. Sept. 1957.
[164] Zoutendijk, G.: Methods of feasible directions. A study in linear and non-linear programming. Amsterdam–London–New York–Princeton, N. J. 1960.
[165] Zurmühl, R.: Matrizen. Darstellung für Ingenieure. 4. print. Berlin–Heidelberg–New York 1964.

ADDENDUM

VERSION OF THE COMPUTER PROGRAMS
FOR PRACTICAL APPLICATION

In line with the textbook nature of the book the programs given in the volume correspond to the principal algorithms. Additions are desirable in practical use for various reasons. A number of letters have indicated there is interest in being able to put the programs directly (without any adaptation) on a particular computer and to use them as they are. On this basis, the authors decided to prepare the present version of the programs for practical application. This version contains, in particular, the additions mentioned in Chapter 3 of the book, namely,

1. The theoretical "tests for zero" are replaced by "epsilon tests" (see page 86 of the book);
2. The printout of the intermediate results was added to the programs (see page 85);
3. Several errors detected so far have been corrected;
4. A number of technical improvements in the programs have been taken into account;
5. The procedures and subroutines have been incorporated into a very simple, ready-to-use driver program.

Since the application version is supposed to allow for the use of the optimization programs without complication, the decomposition algorithm has not been included. On the one hand, that algorithm is not needed for simple cases, and on the other hand, it is efficient only under a balanced utilization of the secondary storage devices.

Details about the use of the programs as well as about peculiarities of the Algol or Fortran version are included directly in the corresponding programs in the form of comments. The reader should consult, in particular, the remarks at the beginning of each program.

ACKNOWLEDGMENT
We should like to express our thanks to Dipl. Ing. Francis Kuhlen for the work of adapting the application version to the computers of the ETH–Zürich and the University of Zürich.

```
begin comment m a t h e m a t i c a l  o p t i m i z a t i o n .
      available methods:        1 standard simplex
               2 dual simplex       3 revised simplex
               5 duoplex          6 gomory — integer
               7 beale — quadratic    8 wolfe — quadratic
      the programs are written for the cdc — 6000 computers of the computing
      center of the swiss federal institute of technology zurich
      (eidgenoessische technische hochschule). algol — 60 — compiler
      version 2.0 with i/o as proposed by d. knuth et al. the use of any
      of the programs requires input in the form of a (repeatable) card
      data deck according to the following specifications:
       — '('6(4d),8d')',method,n,m,m1,m2,m3,eps       (once)
       — '('10( — 4dv3d')')',(a[i,k],k: = 0:n)      (one per row)
      here method is the above method number. the beale and wolfe
      quadratic programs require special input arrangements. the inter —
      mediate tableaus are only printed if each data deck is preceded by
      a card with  p  in column 80. the input value of eps is multiplied
      by 0.0000001


      integer exmpl,m;  boolean stepprint;


procedure inform (a, n, m, p1, p2, zs, ss, ac) ;
      value n,m,zs,ss,ac ; integer n,m,zs,ss,ac ;
      array a ; integer array p1,p2 ;
      begin comment this procedure inform organizes the optional
                  stepwise printoutput of the involved matrices. other output
                  organizations need only a change of this procedure ,
                  inform utilizes global  steppprint ;
      procedure matrixadrucken(a,n,m,p1,p2,zs,ss) ;
            value m,n,zs,ss; integer m,n,zs,ss;
            integer array p1,p2; array a ;
            begin integer i,j,k,l;
                output(61,'('//,'(' linear')',/')');
                for l: = 1 step 10 until n do
                begin j : = if n < l + 9 then n else l + 9;
                      output(61,'('//,18b')');
                      for k: = l step 1 until j do output(61,'('9zd')',k);
                      output(61,'('/,18b')');
                      for k: = l step 1 until j do output(61,'('9zd')',p1[k]);
                      output(61,'('//')');
                      for i: = 0 step 1 until m do
                      begin if i ≠ 0 then
                          output(61,'('b,3zd')',p2[i]) else output(61,'('5b')');
                          if l = 1 then output(61,'(' — 6zd.3d,3b')',a[i*zs])
                              else output(61,'('15b')');
                          for k: = l step 1 until j do
                              output(61,'(' — 4zd.3d')',a[i*zs + k*ss]);
                          output(61,'('/')');
                      end
                end
            end matrixadrucken — matrix a print ;

      procedure matrixbdrucken(b,m,zs,ss) ;
            value m,zs,ss; integer m,zs,ss; array b ;
            begin integer i,j,k,l;
            output(61,'('//,'(' matrix b')',/')');
            for l: = 1 step 10 until m do
                begin j: = if m < l + 9 then m else l + 9 ;
                      output(61,'('//')');
                      for i: = 0 step 1 until m do
                      begin output(61,'('/20b')');
                              for k: = l step 1 until j do
```

5

10

15

20

25

30

35

40

45

55

60

```
                  output(61,'(' − 4zd.3d')',b[i*zs + k*ss])
            end                                                                        65
      end
      output(61,'('/')');
  end matrixbdrucken − matrix b print ;

  procedure matrixcdrucken(c,n,p1);                                                     70
      value n; integer n; integer array p1; array c ;
      begin integer i,j,k,l,m; l: = 2*n + 1;
      output(61,'('//,'(' quadratic objective funtion')'/')');
      for m: = 1 step 10 until n do
            begin j: = if n < m + 9 then n else m + 9;                                  75
            output(61,'('//,18b')');
            for k: = m step 1 until j do output(61,'('9zd')',k);
            output(61,'('/,18b')');
            for k: = m step 1 until j do output(61,'('9zd')',p1[k]);
            output(61,'('//')');                                                        80
            for i: = 0 step 1 until n do
                  begin if i ≠ 0 then output(61,'('b,zzd')',p1[i])
                              else output(61,'('4b')');
                  if m = 1 then output(61,'(' − 4zd.3d,6b')',c[i])
                              else output(61,'('16b')');                                85
                  for k: = m step 1 until j do output(61,'(' − 4zd.3d')',
                        ( if i > k then c[(l − k)*k/2 + i] else c[(l − i)*i/2 + k]));
                  output(61,'('/')');
                  end
            end                                                                        90
      end matrixcdrucken − matrix c print ;
      if ¬stepprint then goto aus ;
      if ac = 1 then matrixadrucken (a,n,m,p1,p2,zs,ss);
      if ac = 2 then matrixbdrucken (a,m,zs,ss);
      if ac = 3 then matrixcdrucken (a,n,p1) ;                                          95
  aus:   end inform ;

  procedure mp1 (a, znr, zschrt, sschrt, l1);
            value znr, zschrt, sschrt; integer znr, zschrt, sschrt;
            array a; integer array l1 ;                                                100
            comment mp1 orders the integers in the list l1[k] such
            that the corresponding real numbers a[znr*zschrt + l1[k]*sschrt]
            decrease monotonically with k;
            begin integer i,k,r,s;
                  for k := 2 step 1 until l 1[0] do                                     105
                        begin i := 0;
                        for s := 1 step 1 until k − 1 do
                              if a[znr*zschrt + l1[k] *sschrt] >
                                    a[znr* zschrt + l1[k − s]* sschrt] then i := i + 1;
                        if i = 0 then goto mp11;                                        110
                        r := l1[k];
                        for s := 1 step 1 until i do l1[k − s + 1]: = l1[k − s];
                        l1[k − i] := r;
  mp11:           end k;
                  end mp1;                                                              115

  procedure mp2 (a,l2,ip,zschrt,sschrt,kp,q1,n,v,eps);
            value kp,zschrt,sschrt,n,v,eps; array a; real q1,eps;
            integer ip, kp, zschrt, sschrt, n, v; integer array l2;
            comment mp2 determines the minimum of all those numbers               120
            v*a[i*zschrt]/a[i*zschrt + kp*sschrt], for which
            a[i*zschrt] ≥ 0 and v*a[i*zschrt + kp*sschrt] > 0, degeneracy is
            taken into consideration;
            begin
            integer i, io, z; real q;                                                  125
            procedure mp4 (a, ip, kp, io, n, zschrt, sschrt, v);
```

```
            value kp, io, n, zschrt, sschrt, v;
            array a; integer ip, kp, io, n, zschrt, sschrt, v;
            comment mp4 handles possible degeneracies;
            begin                                                            130
            integer k; real qp, q0;
            for k : = 1 step 1 until n do
                begin
                qp : = v*a[ip*zschrt + k*sschrt]/a[ip*zschrt + kp*sschrt];
                q0 : = v*a[io*zschrt + k*sschrt]/a[io*zschrt + kp*sschrt];   135
                if qp < q0 then goto mp41;
                if qp > q0 then begin ip : = io; goto mp41 end
                end k;
mp41:       end mp4;
program:    ip : = 0;                                                        140
            for i : = 1 step 1 until l2[0] do
            if v*a[l2[i]*zschrt + kp*sschrt] > eps then
                begin
                q1 : = v* a[l2[i]* zschrt]/ a[l2[i]* zschrt + kp* sschrt];
                q : = q1; ip : = l2[i]; z : = i; goto mp21;                  145
                end
            goto mp22;
mp21:       for i : = z + 1 step 1 until l2[0] do
                begin
                if v*a[l2[i]*zschrt + kp*sschrt] ≤ eps then goto mp211;      150
                q : = v*a[l2[i]*zschrt] / a[l2[i]*zschrt + kp*sschrt];
                if q < q1 then begin ip : = l2[i]; q1 : = q; end
                if q = q1 then
                    begin io : = l2[i];
                    mp4 (a, ip, kp, io, n, zschrt, sschrt, v);              155
                    end
mp211:          end i;
mp22:       end mp2;

procedure mp 3(a, io, i1, ko, k1, ip, kp, zschrt, sschrt, p1, p2);          160
            value io, i1, ko, k1, ip, kp, zschrt, sschrt, p1, p2;
            integer io, i1, ko, k1, ip, kp, zschrt, sschrt, p1, p2;
            array a;
            comment mp3 exchanges a basic and a nonbasic variable. io
            and i1 specify the rows and ko and k1 the columns to which the   165
            transformation is to apply.
            p1 and p2 indicate whether the pivot row und pivot columm,
            respectively, have already been transformed;
            begin
            integer i, k; real piv;                                         170
            piv : = 1/ a[ip*zschrt + kp*sschrt];
            for i : = io step 1 until ip – 1,ip + 1 step 1 until i1 do
                begin
                if p2 = 1 then
                a[i*zschrt + kp*sschrt] : = a[i*zschrt + kp*sschrt]* piv;
                for k: = ko step 1 until kp – 1,kp + 1 step 1 until k1 do    175
                a[i*zschrt + k*sschrt] : = a[i*zschrt + k*sschrt]
                – a[ip*zschrt + k*sschrt]* a[i*zschrt + kp*sschrt];
                end i;
            if p1 = 1 then                                                  180
            for k : = ko step 1 until kp – 1, kp + 1 step 1 until k1 do
            a[ip*zschrt + k*sschrt] : = – a[ip*zschrt + k*sschrt]* piv;
            if p2 = 1 then a[ip*zschrt + kp*sschrt] : = piv;
            end mp 3;
                                                                            185
procedure mp5 (a, l1, q1, kp, znr, zschrt, sschrt);
            value znr, zschrt, sschrt; integer znr, kp, zschrt, sschrt;
            array a; integer array l1; real q1;
            comment mp5 determines the minimum of the numbers
```

```
        a[znr*zschrt + l1[k]*sschrt] with k = 1,2 ... l1[0];               190
        begin
        integer k;
        kp := l1[1]; q1 := a[znr* zschrt + l 1[1]* sschrt];
        for k := 2 step 1 until l1[0] do
        if a[znr* zschrt + l1[k]* sschrt] < q1 then                        195
              begin
              q1 := a[znr* zschrt + l1[k]*sschrt];
              kp := l1[k];
              end
        end mp5;                                                           200

procedure mp7(a, znr, zschrt, sschrt, kp, l1, max);
        value znr, zschrt, sschrt; integer znr, zschrt, sschrt, kp;
        array a; integer array l1; real max;
        comment mp7 determines the maximum of the numbers               205
        a[znr*zschrt + l1[k]*sschrt] with k = 1,2 ... l1[0];
        begin
        integer k;
        max := a[znr* zschrt + l1[1]* sschrt];
        kp := l1[1];                                                      210
        for k := 2 step 1 until l1[0] do
        if a[znr* zschrt + l1[k]* sschrt] > max then
              begin
              max := a[znr* zschrt + l1[k]* sschrt];
              kp := l1[k];                                                215
              end
        end mp7;

procedure mp8 (a, znr, zschrt, sschrt, list, kp, max);
        value znr, zschrt, sschrt; integer znr, zschrt, sschrt, kp ;      220
        real max; array a; integer array list;
        comment determination of max(abs(a[znr*zschrt + list[k]*sschrt]))
        for k = 1,2 ... list[0];
        begin
        integer k;                                                        225
        kp := list[1];
        max := a[znr* zschrt + list[1]* sschrt];
        for k := 2 step 1 until list[0] do
        if abs(max) < abs(a[znr*zschrt + list[k]*sschrt]) then
              begin                                                       230
              kp := list[k];
              max := a[znr* zschrt + list[k]* sschrt];
              end
        end mp8;
                                                                          235
procedure mp9 (a, c, ip, kp, n, m1, zschrt, sschrt, b1);
        value ip, kp, n, m1, zschrt, sschrt, b1;
        integer ip, kp, n, m1, zschrt, sschrt;
        array a, c; boolean b1;
        begin                                                            240
        integer r, s, t, z, z1; real store;
        comment transformation of a bilinear form (used in beale −
             method), all comments refer to fig.16 in the book ;
program:     for r := 0 step 1 until kp − 1,kp + 1 step 1 until n do
              begin z := 0;                                               245
              if r > kp then z1 := kp − 1 else z1 := r;
              for s := 0 step 1 until z1 do
                    begin
                    c[r + z] := c[r + z] + a[ip* zschrt + r* sschrt]* c[kp + z];
                    z := z + n − s;                                       250
                    end s;
              t := z + kp;
```

```
              end r;
         comment the hatched elements have been transformed;
mp91:    for s : = kp + 1 step 1 until n do                                    255
              begin
              for r : = s step 1 until n do
              c[z + n − kp + r]: = c[z + n − kp + r] + a[ip∗zschrt + r∗sschrt]∗c[t + s − kp];
              z : = z + n − s
              end s;                                                           260
         comment from mp91 on the nonhatched elements are transformed;
mp92:    for r : = kp + 1 step 1 until n do
              c[t + r − kp] : = c[t + r − kp] + a[ip∗ zschrt + r∗ sschrt] ∗ c[t];
         comment from mp92 on the elements designated by 1
              are transformed ;                                                265
mp93:    z1 : = 0;
         for r : = 0 step 1 until kp − 1 do
              begin
              z : = 0;
              store : = c[kp + z1] + a[ip∗ zschrt + r∗ sschrt] ∗ c[t];          270
              for s : = 0 step 1 until r do
                   begin
                   c[r + z] : = c[r + z] + a[ip∗ zschrt + s∗ sschrt] ∗ store;
                   z : = z + n − s
                   end s;                                                      275
              z1 : = z1 + n − r;
              end r;
         comment from mp93 on the diagonally hatched elements in the
              upper triangle are transformed for a second time;
mp94:    for r : = kp + 1 step 1 until n do                                    280
              begin z : = 0;
              for s: = 0 step 1 until kp − 1,kp + 1 step 1 until r do
                   begin
                   if s = kp + 1 then z : = z + n − s + 1;
                   c[r + z] : = c[r + z] + a[ip∗ zschrt + s∗ sschrt]∗c[t + r − kp];
                   z : = z + n − s;
                   end s;
              end r;
         comment from mp94 on elements in the rectangle and
              white triangle ;                                                 290
mp95:    if b1 then
              begin z: = 0; comment elements of type 1 and 2;
              for s : = 0 step 1 until kp − 1 do
                   begin
                   c[kp + z] : = c[kp + z] + a[ip∗ zschrt + s∗ sschrt] ∗ c[t];   295
                   z : = z + n − s
                   end s;
              for r: = kp step 1 until n do
              c[r + t − kp] : = c[r + t − kp] + a[ip∗ zschrt + kp∗ sschrt];
              z : = 0;                                                         300
              for s : = 0 step 1 until kp do
                   begin
                   c[kp + z] : = c[kp + z]∗ a[ip∗ zschrt + kp∗ sschrt];
                   z : = z + n − s;
                   end s;                                                      305
              goto mp 97;
              end condition that pivot row in the constraint region;
mp96:    z : = 0;
         for s: = 0 step 1 until kp − 1 do
              begin c[kp + z] : = 0; z: = z + n − s end                        310
         c[t] : = 1 / c[t];
         for r : = kp + 1 step 1 until n do c[r + t − kp] : = 0;
mp97:    end mp9;

procedure simplex (a,zschrt,sschrt,n,m1,m2,m3,fall,proto1,proto2,eps,          315
              inform) ;
```

```
        value zschrt,sschrt,n,m1,m2,m3,eps;  real eps;
        integer zschrt, sschrt, n, m1, m2, m3, fall;
        array a; integer array proto1, proto2;
        procedure inform ;                                                       320
comment routine for optimization of a linear program by the
            simplex method, simplex utilizes global mp2,mp3,mp7,mp8;
        begin
        integer i, k, ip, kp, r, s, v; real max, q1;
        integer array l1[0:n], l2[0:m1 + m2 + m3], l3[0:m2];                      325
        r := 0; v := − 1;
        for k := 1 step 1 until n do l1[k] := proto 1[k] := k;
        l1[0] := n;
        for i := 1 step 1 until m1 + m2 + m3 do l2[i] := i;
        l2[0] := m1 + m2 + m3;                                                   330
        for i := 1 step 1 until m1 + m2 + m3 do proto 2[i]: = n + i;
        if m2 + m3 = 0 then goto s3;
        comment if the origin is a feasible solution, the following
        block up to s3 can be bypassed;
        for i := 1 step 1 until m2 do l3 [i] := 1 ;                               335
        comment computation of the auxiliary objective function for th
            m − method;
soo:    l3[0] := 0;
        r := 1;
        for k := 0 step 1 until n do                                             340
            begin q1 := 0 ;
            for i := m1 + 1 step 1 until m1 + m2 + m3 do
            q1 := q1 + a[i∗ zschrt + k∗ sschrt];
            a[ − zschrt + k∗ sschrt] := − q1;
            end k;                                                               345
        comment computation of a feasible solution by the simplex
            method using the above − computed auxiliary objective function;
s0:     mp7 (a, − 1, zschrt, sschrt, kp, l1, max);
        if max ≤ eps ∧ a[ − zschrt] < − eps then
            begin fall := 2; goto s5 end                                         350
        comment if the maximal coefficient of the auxiliary objective
            function is ≤ 0 and the value of the function is < 0, then
            there exists no feasible solution;
        if max ≤ eps ∧ a[ − zschrt] ≤ eps then
        begin                                                                    355
            for ip := m1 + m2 + 1 step 1 until m1 + m2 + m3 do
                if proto2 [ip] = n + ip then
                begin mp8 (a,ip,zschrt,sschrt,l1,kp,max) ;
                    if max > 0 then goto s01 ;
                end                                                              360
            r := 0 ;
            for i := m1 + 1 step 1 until m1 + m2 do
                if l3 [i − m1] = 1 then
                    for k := 0 step 1 until n do
                        a[i∗zschrt + k∗sschrt]: = − a[i∗zschrt + k∗ sschrt] ;    365
            goto s3 ;
        end if ;
        comment the above condition characterizes a feasible solution;
        mp2 (a,l2,ip,zschrt,sschrt,kp,q1,n,v,eps);
        comment mp2 assures that in the exchange no constraint is             370
            violated;
        if ip = 0 then
            begin fall := 2; goto s5; end
s01:    mp3 (a, − 1, m1 + m2 + m3, 0, n, ip, kp, zschrt, sschrt, 1, 1);
        if proto 2[ip] < n + m1 + m2 + 1 then goto s1;                          375
        for k := 1 step 1 until l1[0] do
        if l1[k] = kp then
            begin
            l1[0] := l1[0] − 1;
```

```
              for s : = k step 1 until l1[0] do l1[s] : =  l1[s + 1];            380
              goto s2;
              end
s1:       if proto2[ip] < n + m1 + 1 then goto s21;
          if l3[proto2[ip] − m1 − n] = 0 then goto s21 ;
          l3 [proto2[ip] − m1 − n] : = 0 ;                                       385
s2:       a[ − zschrt + kp∗ sschrt] : = a[ − zschrt + kp∗ sschrt] + 1;
          for i : = − 1 step 1 until m1 + m2 + m3 do
          a[i∗ zschrt + kp∗ sschrt] : = − a[i∗ zschrt + kp∗ sschrt];
s21:      s : = proto 1[kp];
          proto 1[kp] : = proto 2[ip];
          proto 2[ip]: = s;                                                       390
          inform(a,n,m1 + m2 + m3,proto1,proto2,zschrt,sschrt,1) ;
          if r ≠ 0 then goto s0;
          comment optimization block;
s3:       mp7 (a, 0, zschrt, sschrt, kp, l1, max);                                395
          if max ≤ 0 then begin fall: = 0; goto s5; end
          mp2 (a,l2,ip,zschrt,sschrt,kp,q1,n,v,eps);
s4:       if ip = 0 then begin fall : = 1; goto s5  end
          mp3 (a, 0, m1 + m2 + m3, 0, n, ip, kp, zschrt, sschrt, 1, 1);
          goto s21;                                                               400
s5:       end simplex;


procedure dusex (a,zschrt,sschrt,n,m,fall,w,proto1,proto2,eps,inform);
              value zschrt,sschrt,n,m,w,eps;  integer zschrt,sschrt,n,m,
                    fall;  real eps;                                              405
              array a; integer array proto 1, proto 2; boolean w;
              procedure inform ;
              comment routine for determining the minimum of a linear program
                  by the dual simplex method. the tableau is assumed to be
                  dually feasible. global procedures mp2,mp3,mp5  are used;       410
              begin
              integer v, i, k, ip, kp, znr; real q1;
              integer array l1[0:m], l2[0:n];
program:      for k : = 1 step 1 until m do l1[k]: = k;
              l1[0] : = m;                                                        415
              for i : = 1 step 1 until n do l2[i] : = i;
              l2[0] : = n;
              v : = 1;
              if w then goto dusex 1;
              for k : = 1 step 1 until n do proto 1[k] : = k;
              for i : = 1 step 1 until m do proto 2[i] : = n + i;                 420
dusex1:       i : = zschrt; zschrt : = sschrt; sschrt : = i;
              i : = m; m : = n; n : = i;
              mp5 (a, l1, q1, kp, 0, zschrt, sschrt);
              if q1 ≥ 0 then                                                      425
                    begin fall : = 0 ; goto dusex 3 ; end
              mp2 (a,l2,ip,zschrt,sschrt,kp,q1,n,v,eps);
              if ip = 0 then
                    begin fall : = 2 ; goto dusex 3 ; end
dusex2:       i : = zschrt; zschrt : = sschrt; sschrt : = i;                      430
              i : = m; m : = n; n : = i;
              i : = kp; kp : = ip; ip : = i;
              i : = proto 1[kp]; proto 1[kp] : = proto 2[ip];
              proto 2[ip] : = i;
              mp3 (a, 0, m, 0, n, ip, kp, zschrt, sschrt, 1, 1);                  435
              inform(a,n,m,proto1,proto2,zschrt,sschrt,1) ;
              goto dusex 1;
dusex3:       i : = zschrt; zschrt : = sschrt; sschrt : = i;
              i : = m; m : = n; n : = i;
              end dusex;                                                          440


procedure resex (a,n,m,zschrt,sschrt,fall,proto1,proto2,eps,inform);
```

```
        value n,m,zschrt,sschrt,eps;
        integer n,m,zschrt,sschrt,fall;  real eps;
     array a; integer array proto 1, proto 2;                                445
     procedure inform;
     comment routine for computing the maximum of a linear program
             by the revised simplex method. resex presupposes the origin
             to be a feasible solution. resex uses global mp7;
     begin                                                                   450
     integer i, k, ip, kp, zschrt 1, sschrt 1, s, j;
     real max, q1; array b[m + 1:m∗(m + 2)], c[0:m];
  integer array I1[0:n], I2, I3[0:m] ;
     procedure mp10(b1, c1, m, ip, s, zschrt 2, sschrt 2);
             value m,ip,s,zschrt 2, sschrt 2;                                455
             integer m,ip,s,zschrt 2, sschrt 2; array b1, c1;
             comment routine for the stepwise computation of the
                     inverse matrix occurring in the revised simplex method;
             begin
             integer i, k;                                                   460
             for k : = 1 step 1 until s do
             if b1[ip∗ zschrt 2 + k∗ sschrt 2] ≠ 0 then
                     for i : = 0 step 1 until ip − 1,ip + 1 step 1 until m do
                     if c 1[i] ≠ 0 then
                     b1[i∗zschrt 2 + k∗ sschrt 2] : = b1[i∗ zschrt 2 + k∗ sschrt 2]   465
                     + b1[ip∗ zschrt 2 + k∗ sschrt 2] ∗ c1[i];
                     for k : = 1 step 1 until s do
                       b1[ip∗zschrt2 + k∗sschrt2]: = b1[ip∗zschrt2 + k∗sschrt2]∗c1[ip];
                     end mp10;
                                                                             470
program:  for k: = 1 step 1 until n do proto 1[k] : = I1[k] : = k;
          for i : = 1 step 1 until m do proto 2[i] : = n + i;
          I1[0] : = n; I2[0] : = 0;
          zschrt 1: = 1; sschrt 1: = m + 1;
          comment generation of the identity matrix;                        475
          for i : = 0 step 1 until m do
          for k : = 1 step 1 until m do
          b[i∗ zschrt 1 + k∗ sschrt 1] : = if k = i then − 1 else 0;
          comment determination of the pivot column;
r1:       max : = 0; q1 : = 0;                                               480
          if I1[0] = 0 then goto r2;
          mp7 (a, 0, zschrt, sschrt, kp,I1, max);
r2:       if I2[0] = 0 then goto r3;
          for i : = 1 step 1 until I2[0] do
          if b[0∗ zschrt 1 + I2[i] ∗ sschrt1] > q1 then                      485
                     begin s : = i ;
                     q1 : = b[0∗ zschrt 1 + I2 [i] ∗ sschrt1] ;
                     end
r3:       if max ≤ 0 ∧ q1 ≤ 0 then
                     begin fall : = 0; goto r9 end                          490
          k : = 0;
          if max ≥ q1 then goto r4;
          k : = 1; kp : = I3[s];
          for i : = 0 step 1 until m do
          c[i] : = − b[i∗ zschrt 1 + I2[s] ∗ sschrt1] ;                      495
          goto r5;
r4:       c[0]: = − a[0∗ zschrt + kp∗ sschrt];
          for i : = 1 step 1 until m do
                     begin
                     q1 : = 0;                                              500
                     for j: = 1 step 1 until m do
                     if a[j∗zschrt + kp∗sschrt] ≠ 0 ∧ b[i∗zschrt1 + j∗sschrt1] ≠ 0
                            then
                     q1: = q1 + a[j∗zschrt + kp∗sschrt]∗b[i∗zschrt1 + j∗sschrt1];
                     c[i] : = q1;                                           505
```

```
                    end i;
                comment determination of the pivot row;
r5:             ip : = 0;
                for i : = 1 step 1 until m do
                if c[i] > 0 then                                                 510
                    begin
                    q1 : = a[i* zschrt] / c[i];
                    ip : = i; goto r6;
                    end
                fall : = 1; goto r9;                                             515
r6:             j : = ip;
                for i : = j step 1 until m do
                if c[i] > 0 then
                    begin
                    if a[i* zschrt] / c[i] < q1 then                             520
                        begin q1 : = a[i* zschrt] / c[i]; ip : = i; end
                    end
                for i: = 0 step 1 until ip − 1,ip + 1 step 1 until m do
                    c[i] : = − c[i] / c[ip];
                c[ip] : = 1 / c[ip];                                             525
                comment transformation of the first column of tableau a;
                mp 10 (a, c, m, ip, 1, zschrt, 0);
                comment transformation of the first row of tableau a;
                for j : = 1 step 1 until kp − 1, kp + 1 step 1 until n do
                    begin q1 : = 0 ;                                             530
                    for i : = 1 step 1 until m do
                    if b[ip* zschrt 1 + i* sschrt1] ≠ 0 then
                    q1 : = q1 + b[ip*zschrt1 + i*sschrt1] * a[i*zschrt + j*sschrt];
                    a[0* zschrt + j* sschrt] : = a[0* zschrt + j* sschrt] − q1 * c[0];
                    end j;                                                       535
                a[0* zschrt + kp* sschrt] : = − c[0];
                comment transformation of the inverse;
                mp 10 (b, c, m, ip, m, zschrt 1, sschrt 1);
                if k = 0 then
                    begin                                                       540
                    for j : = 1 step 1 until l2[0] do
                    if ip = l2[j] then
                        begin
                        if l3[j] = kp then goto r8;
                        for i : = 1 step 1 until l1[0] do                       545
                        if l1[i] = kp then l1[i] : = l3[j];
                        l3[j] : = kp; goto r8;
                        end
                    l2[0] : = l2[0] + 1; l2[l2[0]] : = ip; l3[l2[0]] : = kp;
r7:                 for i : = 1 step 1 until l1[0] do
                    if l1[i] = kp then
                        begin l1[0] : = l1[0] − 1;
                        for j : = i step 1 until l1[0] do l1[j] : = l1[j + 1];
                        goto r8;
                        end for if;                                             555
                    end if;
                    for j : = 1 step 1 until l2[0] do
                    if ip = l2[j] then goto r8;
                    l1[0] : = l1[0] + 1; l1[l1[0]]: = l3[s] ; l2 [0] : = l2[0] − 1 ;
                    for j : = s step 1 until l2[0] do                           560
                        begin l2[j] : = l2[j + 1] ; l3[j] : = l3[j + 1] end
r8:             k : = proto 1[kp]; proto 1[kp] : = proto 2[ip]; proto 2[ip] : = k;
                inform(a,n,m,proto1,proto2,zschrt,sschrt,1) ;
                inform(b,n,m,proto1,proto2,zschrt1,sschrt1,2);
                goto r1;                                                        565
r9:         end resex;
procedure duoplex (a,n,m1,m2,zschrt,sschrt,fall,proto1,proto2,eps,
            inform) ;
```

```
          value n,m1,m2,zschrt,sschrt,eps;  real eps;                          570
          integer n, m1, m2, zschrt, sschrt, fall;
          array a; integer array proto 1, proto 2;
          procedure inform ;
          comment routine to determine the maximum of a linear program
                  by the duoplex method. duoplex uses global procedures mp1,  575
                  mp2,mp3,mp5;
       begin
          integer v, i, k, ip, kp, r, s, z, j; real q1, q2, q3;
          integer array l1[0:n], l2[0:m1 + m2], l3[0:m2];
          procedure list (m1,m2,a,l2, z, zschrt);                             580
             value m1, m2, zschrt;
             integer m1, m2, z, zschrt;
             array a ; integer array l2;
             comment list inserts all row indices  i  with a[i*zschrt]
                ≥ 0, i ≥ 1, into list l2 and associates  z  with the index   585
             of the first row for which a[i*zchrt] < 0;
             begin
             integer i;
             l2[0] : = z : = 0;
             for i : = 1 step 1 until m1 + m2 do                             590
                begin
                if a[i* zschrt]≥ 0 then
                   begin l2[0] : = l2[0] + 1;
                      l2[l2[0]] : = i; goto list1;
                      end                                                    595
                   if z = 0 then z : = i;
list1 :         end i;
             end list;
program:     l1[0]: = n; j: = 1; r: = 0;
          for k : = 1 step 1 until n do l1[k] : = proto 1[k] : = k;          600
          for i : = 1 step 1 until m1 + m2 do proto 2[i] : = n + i;
          if m2 = 0 then begin l3[0] : = 0; goto mw; end
          comment if there are no equality constraints, the next block
                  up to mw is bypassed;
equations:   l3[0]: = m2;                                                    605
          for i : = 1 step 1 until m2 do l3[i] : =. m1 + i;
          comment variables belonging to the equations are removed from
          the basis by means of the multiphase method;
gl1:      list (m1, m2, a, l2, z, zschrt);
          mp5 (a, l1, q1, kp, l3[1], zschrt, sschrt);                        610
          if q1 ≥ 0 then
             begin fall : = 2;  goto duoplex1;  end
          mp2 (a,l2,ip,zschrt,sschrt,kp,q1,n, − 1,eps);
gl2:      if ip = 0 then
             begin                                                          615
             ip : = l3[1];
             l3[0] : = l3[0] − 1;
             for s : = 1 step 1 until l3[0] do
                l3[s] : = l3[s + 1];
             l1[0] : = l1[0] − 1;                                           620
             for s: = 1 step 1 until l1[0] do if l1[s] = kp then
                for k: = s step 1 until l1[0] do l1[k]: = l1[ k + 1];
             goto gl 3;
             end
          q2 : = a[l3[1]* zschrt]/( − a[l3[1]* zschrt + kp* sschrt]);        625
          if q2 ≤ q1 then
             begin ip : = 0;  goto gl2;  end
          if ip ≤ m1 then goto gl3;
          for s : = 1 step 1 until l3[0] do
          if ip = l3[s] then                                                630
             begin
             l3[0] : = l3[0] − 1; l1[0] : = l1[0] − 1;
```

```
              for k: = s step 1 until I3[0] do I3[k] : =  I3[k + 1];
              end
          for s: = 1 step 1 until I1[0] do if I1[s] = kp then                    635
              for k: = s step 1 until I1[0] do I1[k]: =  I1[ k + 1];
gl3:      s : = proto 1[kp];
          proto 1[kp]: =  proto 2[ip];
          proto 2[ip] : =  s;
          mp3 (a, 0, m1 + m2, 0, n, ip, kp, zschrt, sschrt, 1, 1);          640
          inform(a,n,m,proto1,proto2,zschrt,sschrt,1) ;
          if I3[0] = 0 then goto mw;
          goto gl1;
          comment in the block beginning with mw that constraint is
          determined for which the angle between its normal and the          645
          gradient of the objective function is largest;
mw:       list (m1, m2, a, I2, z, zschrt);
          mp1 (a, 0, zschrt, sschrt, I1);
          kp : =  I1[1];
          if a[I1[1]* sschrt] ≤ 0 ∧ z = 0 then                              650
              begin
              if r = 0 ∨ a[r* sschrt] ≤ 0 then
                  begin fall : = 0; goto duplex 1;  end
              I1[0] : =  I1[0] + 1; I1[I1[0]] : =  r; kp: =  r; r: = 0; goto s2;
              end
          if j = 0 then goto s1;                                             655
          q2 : =  ₁₀30;
          for i : = 1 step 1 until m1 + m2 do
              begin
              q1 : =  q3 : =  0;                                            660
              for k : = 1 step 1 until I1[0] do
                  begin
                  q1 : =  q1 + a[I1[k]*sschrt]*a[i*zschrt + I1[k]*sschrt];
                  q3 : =  q3 + a[i* zschrt + I1[k]* sschrt]↑2;
                  end k;                                                     665
              if q1 ≥ 0 ∧ q2 < 0 then goto mw1;
              if q1 < 0 then q1 : =  − q1 ↑2 else q1 : =  q1 ↑2;
              q1 : =  q1/q3;
              if q1 < q2 then begin q2 : =  q1; ip : =  i; end
mw1:          end i;                                                         670
          k : = 1;
mw2:      if abs(a[ip*zschrt + I1[k]*sschrt]) > eps then
              begin
              r : = kp : =  I1[k];
              I1[0] : =  I1[0] − 1;                                         675
              for s : = k step 1 until I1[0] do I1[s] : =  I1[s + 1];
              j : = 0;
              goto gl3;
              end
          k : =  k + 1;                                                     680
          goto mw2;
          comment in the next block the still violated constraints are
          being satisfied;
s1:       if z = 0 then goto s2;
          mp1 (a, z, zschrt, sschrt, I1);                                   685
          if a[z* zschrt + I1[1]* sschrt] ≤ 0 then
              begin
              if a[z* zschrt + r* sschrt] ≤ 0 then
                  begin fall : = 2;  goto duplex 1;  end
              kp : = r; I1[0] : =  I1[0] + 1; I1[I1[0]] : =  r; r : = 0; goto s2;   690
              end
          kp : =  I1[1];
s2:       mp2 (a, I2, ip, zschrt, sschrt, kp, q1, n, − 1, eps) ;
          if ip = 0 then
              begin                                                         695
```

```
            if z = 0 then
                    begin fall : = 1;  goto duoplex 1;  end
            ip : = z;
            end
         goto gl3;                                                              700
duoplex1:      end duoplex;

procedure gomory (a,n,m1,m2,m3,zschrt,sschrt,fall,proto1,proto2,eps,
            inform,m);
            value n,m1,m2,m3,zschrt,sschrt,eps,m;  real eps;                    705
            integer n, m1, m2, m3, zschrt, sschrt, fall;  array a;
            integer array proto 1, proto 2;  integer m;
            procedure inform ;
            comment routine for the minimization of a linear program under
                    the additional condition that all independent variables are    710
                    integer — valued.  gomory uses simplex and dusex globally. the
                    tableau is stored row by row ,
                    m is the declared maximal number of rows ;
            begin
            integer i, k, j, m4;  boolean w;                                    715
g01:        for k : = 0 step 1 until n do
            a[0* zschrt + k* sschrt] : =  — a[0* zschrt + k* sschrt];
            simplex (a,zschrt,sschrt,n,m1,m2,m3,fall,proto1,proto2,eps,
                    inform) ;
            if fall > 0 then goto g04;                                          720
            output(61,'('//'(' non — integer optimum found')',/')');
            m4 : = m1 + m2 + m3 ;
            j : = n;
            for k : = 1 step 1 until n — m3 do
g11:            if proto1[k] > n + m1 + m2 ∧ proto1[k] ≤ n + m4 then           725
                    begin proto1[k] : = proto1[j];
                    for i : = 0 step 1 until m4 do
                        a[i*zschrt + k*sschrt] : = a[i*zschrt + j*sschrt];
                    j : = j — 1;  goto g11;
                    end k;   n : = n — m3;                                      730
            for k : = 0 step 1 until n do
            a[0* zschrt + k* sschrt] : =  — a[0* zschrt + k* sschrt];
g02:        for i : = 1 step 1 until m4 do
            if proto 2[i] ≤ n ∧ abs(entier(a[i* zschrt] + .1) — a[i*zschrt])
                    > eps then                                                 735
                    begin
                    if m4 ≥ m then
                        begin output(61,'('//'(' ****numerically instable')'
                            ,/')');  goto g04 end
                    m4 : = m4 + 1;  proto 2[m4] : = n + m4; w : = true         740
                    a[m4* zschrt] : = entier (a[i* zschrt]) — a[i* zschrt];
                    for k : = 1 step 1 until n do
                    a[m4* zschrt + k* sschrt] : =  — a[i* zschrt + k* sschrt] —
                    entier( — a[i* zschrt + k* sschrt]);
                    goto g03;                                                  745
                    end
            goto g04;
g03:        dusex (a,zschrt,sschrt,n,m4,fall,w,proto1,proto2,eps,inform);
            if fall > 0 then goto g04;
            goto g02;                                                          750
g04:        end gomory;

procedure beale (c,a,n,m1,zschrt,sschrt,proto1,proto2,fall,eps,inform);
            value n,zschrt,sschrt,eps;  integer n,m1,zschrt,sschrt,fall;
            array c,a;  integer array proto1,proto2;  real eps;                755
            procedure inform ;
            comment routine for determining the minimum of a definite
                    quadratic form subject to linear constraints. beale uses
```

```
               global procedures mp2,mp3,mp5,mp8,mp9;
               begin                                                              760
               integer i,k,ip,kp,v,s,z,m ,t,r;  real max,q1;
               integer array l1[0:n], l2[0:n + m1], list[0:n], ablist[1:n];
               boolean b1;
program:       m : = m1 ;
               for k: = 1 step 1 until n do l1[k]: = proto1[k]: = k;              765
               for i : = 1 step 1 until m do l2[i] : = i;
               for i : = 1 step 1 until m do proto 2[i]: = n + i;
               l1[0] : = n; l2[0] : = m;
               for k : = 1 step 1 until n do ablist[k] : = 0;
               list[0] : = 0;                                                     770
               comment the block beale 1 determines the pivot column kp;
beale1:        if list[0] ≠ 0 then
                       begin
                       mp8 (c, 0, 0, 1, list, kp, max);
                       if abs(max) > eps then goto beale 2;                       775
                       end
               if l1[0] > 0 then mp5 (c, l1, q1, kp, 0, 0, 1) else q1 : = 1;
               if q1 ≥ − eps then
                       begin fall : = 0;  goto beale 6; end
               comment if all leading elements of u − columns = 0 and no         780
               leading element of an x − column < 0, the minimum has been
               reached;
               comment the block reale 2 determines the pivot row;
beale2:        max : = if c[kp* (n + 1) − (kp* (kp − 1))/2] > eps then
                       c[kp] / c[kp* (n + 1) − (kp* (kp − 1))/2] else 0.0;        785
               v : = if max > 0 then 1 else − 1;
               mp2 (a,l2,ip,zschrt,sschrt,kp,q1,n,v,eps);
               if ip = 0 ∧ max = 0 then
                       begin fall : = 1; goto beale 6; end
               comment if ip = 0 and max = 0, the solution is unbounded ;        790
               if ip = 0 then goto beale 4;
beale3:        if q1 ≤ abs(max) then
                       begin comment pivot row in the constraint section;
                       i: = proto2[ip]; proto2[ip]: = proto1[kp]; proto1[kp]: = i;
                       mp3 (a, 1, m, 0, n, ip, kp, zschrt, sschrt, 1, 1);          795
                       mp9 (a,c,ip,kp,n,m1,zschrt,sschrt, true );
                       if proto2[ip] > n + m1 then
                               begin proto2[ip] : = proto2[m]; l2[0] : = l2[0] − 1;
                               l1[0]: = l1[0] + 1; l1[l1[0]]: = kp; t: = list[0];list[0]: = t − 1;
                               for k: = 1 step 1 until list[0] do                  800
                                       if list[k] = kp then t : = k;
                               for k: = t step 1 until list[0] do list[k]: = list[k + 1];
                               for k: = 0 step 1 until n do
                                       a[ip*zschrt + k*sschrt] : = a[m*zschrt + k*sschrt];
                               m : = m − 1; ablist[kp] : = ablist[kp] − 1;         805
                               end if;
                       inform(c,n,m,proto1,proto2,zschrt,sschrt,3);
                       inform(a,n,m,proto1,proto2,zschrt,sschrt,1);
                       goto beale 1;
                       end                                                         810
beale4:        ablist[kp] : = ablist[kp] + 1; t : = 0; r : = 0;
beale4a:       for i: = 1 step 1 until list[0] do
               if list[i] = kp then
                       begin z: = 0; comment u − column exchange;
                       for s : = 0 step 1 until kp do                             815
                               begin a[t*zschrt + s*sschrt] : = c[kp + z];
                               z : = z + n − s;
                               end s;
                       k : = kp*(n + 1) − kp*(kp − 1)/2; ip : = t;
                       for s: = 1 step 1 until n − kp do                          820
                               a[t*zschrt + (kp + s)*sschrt] : = c[k + s];
```

```
            mp3 (a, r, m, 0, n, ip,kp, zschrt, sschrt, 1, 1);
            mp9 (a,c,ip,kp,n,m1,zschrt,sschrt, false );
            inform(c,n,m,proto1,proto2,zschrt,sschrt,3);
            inform(a,n,m,proto1,proto2,zschrt,sschrt,1);                    825
            goto beale 1;
            end
beale 5:    list[0] : = list[0] + 1;
            list[list[0]] : = kp;
            proto 2[m + 1] : = proto 1[kp];                                 830
            proto 1[kp] : = n + m1 + kp;
            l2[m + 1] : = l2[0] : = m + 1;
            for s : = 1 step 1 until l1[0] do if l1[s] = kp then
                        t : = s;
            l1[0] : = l1[0] − 1;                                            835
            for s : = t step 1 until l1[0] do l1[s] : = l1[s + 1];
            m : = t : = m + 1; r : = 1;
            goto beale 4a;
beale 6:      m1 : = m ;
            end beale ;                                                     840


procedure wolfe (a,x,zschrt,sschrt,n,m,proto1,proto2,fall,eps,inform);
            value zschrt,sschrt,n,m,eps;  real eps;
            integer zschrt, sschrt, n, m, fall;
            array a, x; integer array proto 1, proto 2;                     845
            procedure inform ;
            comment routine for computing the minimum of a quadratic form
                    subject to linear constraints by means of the long form of
                    wolfe − method.  uses global procedures mp2,mp3,mp5 ;
            begin                                                           850
            integer i, k, ip, kp, s, t, n1, v;
            integer array l1[0:3*n + m + 1], l2[0:m + n], list1[0:m + n + 1];
            real q1, q2;
procedure mp 11 (a, zschrt, sschrt, n, m, list 1);
            value m, zschrt, sschrt; integer n, m, zschrt, sschrt;         855
            array a; integer array list 1;
            comment procedure for removal of the columns of tableau a
                    contained in list 1;
            begin
            integer i,k,s,t;  t : = 0;                                      860
            for k : = 1 step 1 until list 1[0] do
            begin
                    for s : = list 1[k] step 1 until n − 1 do
                    for i : = 0 step 1 until m do
                    a[i*zschrt + (s − t)* sschrt] : = a[i* zschrt + (s − t + 1)* sschrt];    865
                    t : = t + 1;
            end k;
            n : = n − list 1[0];
            end mp 11;
procedure mp 12 (proto 1, proto 2, l1, m, n, n1);                          870
            value m, n, n1; integer m, n, n1;
            integer array proto 1, proto 2, l1;
            comment listing of all columns admissible as pivot columns;
            begin integer k,i;  l1[0] : = 0;
            for k : = 1 step 1 until n1 do                                  875
                    begin
                    if proto 1[k] > n then goto mp121;
                    for i : = 1 step 1 until m + n do
                    if proto 1[k] + n = proto 2[i] then goto mp 123;
mp 122:             l1[0] : = l1[0] + 1; l1[l1[0]] : = k;                   880
                    goto mp 123;
mp 121:             if proto1[k] > 2*n then goto mp 122;
                    for i : = 1 step 1 until m + n do
                    if proto 1[k] − n = proto 2[i] then goto mp 123;
```

```
mp123:              goto mp 122;                                           885
                    end k;
                 end mp 12;

                 comment initialization of the indices and index lists;
program:         for k := 1 step 1 until 3 * n + m + 1 do                  890
                 proto 1[k] := k;
                 for i: = 1 step 1 until m + n do proto2[i] := 3*n + m + 1 + i;
                 for k: = 1 step 1 until n, n + 1 + m + 1 step 1 until 3*n + m do
                 if k ≤ n then l1[k] := k else l1[k − n − m] := k;
                 l1[0] := 2 * n;                                           895
                 for i: = 1 step 1 until m + n do l2[i] := i;
                 l2[0] := m + n;
                 v := − 1; n1 := 3 * n + m + 1;
                 comment first phase of wolfe − method.
                 part 1a, routine for computing the auxiliary objective function;   900
w1:              for k : = 0 step 1 until n do
                    begin q1 := 0;
                    for i : = 1 step 1 until m do
                    q1 := q1 + a[i* zschrt + k* sschrt];
                    a[0* zschrt + k* sschrt] := q1;                        905
                    end k;
                 for k: = n + 1 step 1 until 3*n + m + 1 do a[0*zschrt + k*sschrt]: = 0;
                 comment 1b, minimization of the auxiliary objective function
                 by the simplex method;
                 output(61,'('//'('( wolfe 1')'/')');                     910
w11:             mp5 (a, l1, q1, kp, 0, zschrt, sschrt);
                 if q1 ≥ 0 then goto w 12;
                 comment if q1 ≥ 0, the minimum of the auxiliary objective
                 function has been obtained;
                 mp2 (a,l2,ip,zschrt,sschrt,kp,q1,3*n + m + 1,v,eps);     915
                 if ip = 0 then begin fall := 2; goto w4; end
                 i : = proto1[kp]; proto1[kp] := proto2[ip]; proto2[ip] := i;
                 mp3 (a,0,m + n,0,3*n + m + 1,ip,kp,zschrt,sschrt,1,1);
                 inform(a,zschrt − 1,m + n,proto1,proto2,zschrt,sschrt,1);
                 goto w 11;                                                920
                 comment 1c. deletion of the w − and z − columns not contained
                 in the basis;
w12:             list 1[0] := 0; i := 0;
                 for k : = 1 step 1 until n1 − 1 do
                    begin                                                  925
w121:               if k > n1 − i − 1 then goto w 122;
                    if proto1[k] ≥ 2*n + m + 1 ∧ proto1[k] ≤ 3*n + m ∨
                         proto1[k] ≥ 3*n + m + 2 ∧ proto1[k] ≤ 4*n + 2*m + 1 then
                         begin list 1[0] := list 1[0] + 1;
                         list 1[list 1[0]] := k + i;                       930
                         for s: = k step 1 until n1 − i − 1 do
                              proto1[s] := proto1[s + 1];
                         i := i + 1; goto w 121;
                         end if;
                    end k;                                                 935
w122:            mp11 (a,zschrt,sschrt,n1,m + n,list 1);
                 output(61,'('//'('( wolfe 2')'/')');
                 inform(a,n1,m + n,proto1,proto2,zschrt,sschrt,1);
                 comment start of the second phase of wolfe − method.
                 part 2a. computation of the second auxiliary objective   940
                 function;
w2:              for k : = 0 step 1 until n1 do
                    begin q1 := 0;
                    for i : = 1 step 1 until n + m do
                    if proto 2[i] ≥ 2 * n + m + 1 ∧ proto 2[i] ≤ 3 * n + m   945
                         ∨ proto2[i] ≥ 3*n + 2*m + 2 ∧ proto2[i] ≤ 4*n + 2*m + 1 then
                    q1 := q1 + a[i* zschrt + k* sschrt];
```

```
              a[0*zschrt + k* sschrt] : = q1;
              end k;
          comment 2b. minimization of the second auxiliary objective                950
          function;
w21:      mp12 (proto 1, proto 2, l1, m, n, n1 − 1);
          inform(a,n1,m + n,proto1,proto2,zschrt,sschrt,1);
          mp5 (a, l1, q1, kp, 0, zschrt, sschrt);
          if q1 ≥ 0 then goto w22;                                                   955
          mp2 (a,l2,ip,zschrt,sschrt,kp,q1,n1,v,eps);
          if ip = 0 then begin fall : = 2; goto w4; end
          i : = proto1[kp]; proto1[kp] : = proto2[ip]; proto2[ip] : = i;
          mp3 (a, 0, m + n, 0, n1, ip, kp, zschrt, sschrt, 1, 1);
          goto w 21;                                                                 960
          comment 2c. deletion of the z − columns from the tableau;
w22:      i : = list 1[0] : = 0;
          for k : = 1 step 1 until n1 − 1 do
              begin
w221:         if k > n1 − 1 − i then goto w 222;                                     965
              if proto1[k] ≥ 2*n + m + 1 ∧ proto1[k] ≤ 3*n + m ∨
                    proto1[k] ≥ 3*n + 2*m + 2 ∧ proto1[k] ≤ 4*n + 2*m + 1 then
                  begin list1[0]: = list1[0] + 1; list1[list1[0]] : = k + i;
                  for s: = k step 1 until n1 − 1 − i do proto1[s]: = proto1[s + 1];
                  i : = i + 1; goto w 221;                                           970
                  end if;
              end k;
w222:     mp11 (a,zschrt,sschrt,n1,m + n,list 1);
          output(61,'('//'(' wolfe 3')'/')');
          inform(a,n1,m + n,proto1,proto2,zschrt,sschrt,1);                          975
          for k: = 0 step 1 until n1 − 1 do a[0*zschrt + k*sschrt]: = 0;
          a[0* zschrt + n1 * sschrt] : = − 1;
          comment part 3. start of the third phase of wolfe − method;
w3:       mp12 (proto 1, proto 2, l1, m, n, n1);
          x[0] : = a[0* zschrt + 0* sschrt];                                         980
          for i: = 1 step 1 until m + n do x[i]: = 0;
          for i: = 1 step 1 until m + n do if proto2[i] ≤ n then
          x[proto 2[i]] : = a[i* zschrt + 0* sschrt];
          mp5 (a, l1, q1, kp, 0, zschrt, sschrt);
          if q1 ≥ 0 then begin fall : = 2; goto w4; end                             985
          mp2 (a,l2,ip,zschrt,sschrt,kp,q1,n1,v,eps);
          if ip = 0 then begin fall : = 2; goto w4; end
          i : = proto1[kp]; proto1[kp] : = proto2[ip]; proto2[ip] : = i;
          mp3 (a, 0, m + n, 0, n1, ip, kp, zschrt, sschrt, 1, 1);
          inform(a,n1,m + n,proto1,proto2,zschrt,sschrt,1);                          990
          if − a[0* zschrt + 0* sschrt] < 1 then goto w3;
          comment computation of the solution;
w33:      q1: = ( − a[0] − 1)/( − a[0] + x[0]); q2: = (1 + x[0])/( − a[0] + x[0]);
          for i: = 1 step 1 until n do x[i] : = q1 * x[i];
          for i: = 1 step 1 until m + n do if proto2[i] ≤ n then                     995
              x[proto2[i]] : = x[proto2[i]] + q2 * a[i* zschrt];
          fall : = 0; proto1[0] : = n1;
w4:       end wolfe;

          integer i,k,m1,m2,m3,n,method,zs,ss,fall,zt; real r,eps;                  1000
          stepprint : = false ; exmpl : = 1;
l1:       input(60,'('6(4d),8d,47b,a,/')',method,n,m,m1,m2,m3,zt,k);
          eps : = zt * 0.0000001 ;
          if k ≠ equiv('(' ')') then begin stepprint : = k = equiv('('p')');
              goto l1 end                                                            1005
          if method = 0 then goto omega;
          output(61,'(' ↑,'(' mathematical optimization  ')',zd,20b,
              '('example no.')',zzd,//')',method,exmpl);
          output(61,'(''('  n = ')',zd,'('  m = ')',zd,'(' m1 = ')',zd,
              '(' m2 = ')',zd,'(' m3 = ')',zd,//')',n,m,m1,m2,m3);                  1010
```

```
output(61,'(' '(' epsilon = ')', - d.8d ₁₀ + dd,//')',eps);
k : = m1 + m2 + m3;
if k > 0 ∧ k ≠ m then
        output(61,'(' '(' **** constraint number m ≠ m1 + m2 + m3 ')',//')');
zs : = n + 1; ss : = 1;                                                            1015
if method = 7 then goto mbeale;
if method = 8 then goto mwolfe;
        begin
        array a[ − n − 1 : (m + 1)*(n + 1)];
        integer array proto1[0:n], proto2[0:m];                                    1020
        for i : = 0 step 1 until m do
                begin proto2[i] : = n + i;
                for k : = 0 step 1 until n do
                        input(60,'(' − 4dv3d')',a[i*zs + k*ss]);
                input(60,'('/')');                                                 1025
                end i;
        for k : = 0 step 1 until n do proto1[k] : = k;
        proto2[0] : = 0; fall : = 9;
        inform(a,n,m,proto1,proto2,zs,ss,1);
        if method = 1 then begin                                                   1030
                output(61,'('//'(' standard simplex method')',/')');
                simplex (a,zs,ss,n,m1,m2,m3,fall,proto1,proto2,eps,inform);
                end
        if method = 2 then begin
                output(61,'('//'(' dual standard simplex method')'/')');           1035
                dusex(a,zs,ss,n,m,fall, false ,proto1,proto2,eps,inform);
                end
        if method = 3 then begin
                output(61,'('//'(' revised simplex method')'/')');
                resex(a,n,m,zs,ss,fall,proto1,proto2,eps,inform);                  1040
                end
        if method = 5 then begin
                output(61,'('//'(' standard duoplex method')'/')');
                duoplex(a,n,m1,m2,zs,ss,fall,proto1,proto2,eps,inform);
                end                                                                1045
        if method = 6 then begin
                output(61,'('//'(' integer algorithm of gomory')'/')');
                gomory(a,n,m1,m2,m3,zs,ss,fall,proto1,proto2,eps,inform,m);
                end
        if fall = 0 then                                                          1050
                output(61,'('//'(' finite solution')'/')');
        if fall = 1 then
                output(61,'('//'(' solution unbounded')'/')');  ·
        if fall = 2 then
                output(61,'('//'(' no feasible solution')'/')');                  1055
        inform(a,n,m,proto1,proto2,zs,ss,1);
        exmpl : = exmpl + 1; goto l1;
        end array − block;
mbeale:         begin
        array a[0:(m + n + 2)*(n + 1)], c[0:n*(n + 3)/2];                          1060
        integer array proto1[0:n], proto2[0:m + n];
        output(61,'('//'(' quadratic optimization of beale')'/')');
        proto2[0] : = 0 ;
        for i: = 0 step 1 until n do
        begin proto1[i] : = i ; a[i] : = 0 ;                                       1065
          for k : = 0 step 1 until i − 1 do input(60,'(' − 4dv3d')',r);
          for k : = i step 1 until n do input(60,'(' − 4dv3d')',
                        c[(n + n − i + 1)*i/2 + k]);
          input(60,'('/')');
        end input of quadratic objective function ;                               1070
        inform(c,n,m,proto1,proto2,zs,ss,3);
        for i: = 1 step 1 until m do
        begin proto2[i] : = n + i;
```

```
        for k : = 0 step 1 until n do input(60, '(' − 4dv3d')',
                        a[i*zs + k*ss]);                                          1075
        input(60, '('/')');
        end input of linear constraints ;
        inform(a,n,m,proto1,proto2,zs,ss,1);
        beale(c,a,n,m,zs,ss,proto1,proto2,fall,eps,inform);
        if fall = 0 then                                                          1080
            output(61, '('//'(' finite solution')'/')');
        if fall = 1 then
            output(61, '('//'(' solution unbounded')'/')');
        inform(c,n,m,proto1,proto2,zs,ss,3);
        inform(a,n,m,proto1,proto2,zs,ss,1);                                      1085
        exmpl : = exmpl + 1; goto l1;
        end beale;
mwolfe:  begin
        array a[0:(3*n + m + 2)*(m + n + 1)], x[0:4*n + 2*m + 1];
        integer array proto1[0:3*n + m + 1], proto2[0:m + n];                     1090
        zs : = 3*n + m + 2; ss : = 1; m2 : = m + n + 1;  proto2[0] : = 0;
        output(61, '('//'(' quadratic optimization of wolfe')'/')');
        for i : = 0 step 1 until zs*m2 do a[i] : = 0;
        for i : = 1 step 1 until m + n do
            begin proto2[i] : = zs − 1 + i;                                       1095
            for k: = 0 step 1 until n do input(60, '(' − 4dv3d')',
                    a[i*zs + k*ss]);
            input(60, '('/')');
            end i;
        for k: = 0 step 1 until zs − 1 do proto1[k] : = k;                        1100
        inform(a,n,m + n,proto1,proto2,zs,ss,1);
        for i: = 1 step 1 until n do
            begin m3 : = (m + i)*zs; a[m3 + (n + i)*ss] : = a[m3 + (m + n + n + i)*ss]
                : = 1; a[m3 + zs − 1] : = a[m3]; a[m3] : = 0;
            for k: = 1 step 1 until m do                                          1105
                a[m3 + (n + n + k)*ss] : = a[k*zs + i*ss];
            end i;
        inform(a,zs − 1,m + n,proto1,proto2,zs,ss,1);
        wolfe (a,x,zs,ss,n,m,proto1,proto2,fall,eps,inform);
        if fall = 1 then                                                         1110
            output(61, '('//'(' solution unbounded')'/')');
        if fall = 2 then
            output(61, '('//'(' no feasible solution')'/')');
        exmpl : = exmpl + 1;  if fall > 0 then goto l1;
        output(61, '('//'(' finite solution')'/')');                            1115
        inform(a,zs − 1,m + n,proto1,proto2,zs,ss,1);
        output(61, '('/'(' x − values of the original independent ')',
                '('variables')'/')');
        for i: = 1 step 1 until n do
            begin if i//10*10 = i − 1 then output(61, '('/,20b')');              1120
                output(61, '(' − 4zd.3d')',x[i]) end
        goto l1;
        end wolfe ;
omega:
end driver program, mathematical optimization ;                                  1125
            eop
            finis
```

```
         PROGRAM MATOPT (INPUT,OUTPUT)
C        M A T H E M A T I C A L   O P T I M I Z A T I O N
C        FOR THE MAIN COMMENTS SEE INTRODUCTORY COMMENTS IN THE ALGOL –
C        PROGRAMS.
C        FORTRAN – RUN – COMPILER VERSION 2.3 , PROGRAM MODE : FORTRAN IV          5
C        INPUT: (6I4,I8),METHOD,N,M,M1,M2,M3,EPS          (ONCE)
C           (10F8.3) FOR  (A(I,J),J = 1,N1)  WITH N1 = N + 1   (ONE PER ROW)
C        INTERMEDIATE TABLEAUS ARE ALWAYS PRINTED.
         DIMENSION A(1000),B(1000),C(1000),X(200),L1(50),L2(50),L3(50),
       1    LIST(50),ABLIST(50),LIST1(50)
         INTEGER PROTO1(200),PROTO2(200),MET,SS,ZS,FALL,D,DA,DB,DC,EXMPL,ZT    10
         EXMPL = 1
900      READ 901,MET,N,M,M1,M2,M3,ZT
901      FORMAT(6I4,I8,47X,R1)
         IF(MET.EQ.0) GO TO 999                                                  15
         EPS = 0.0000001*ZT
         PRINT 906,MET,EXMPL
906      FORMAT(28H1MATHEMATICAL OPTIMIZATION   ,I2,20X,12HEXAMPLE NO. ,I3)
         PRINT 907,N,M,M1,M2,M3
907      FORMAT(3H0N = ,I2,5H   M = ,I2,4H M1 = ,I2,4H M2 = ,I2,4H M3 = ,I2,/)   20
         PRINT 908,EPS
908      FORMAT(11H EPSILON  = ,E15.2,/)
         MK = M1 + M2 + M3
         IF(MK.GT.0.AND.MK.NE.M) PRINT 1001
1001     FORMAT(* $$$$CONSTRAINT NUMBER M ≠ M1 + M2 + M3*/)                       25
         SS = 1
         ZS = N + 1
         MA = M + 1
         IF(MET.EQ.7) GO TO 700
         IF(MET.EQ.8) GO TO 800                                                   30
         DO 912 I = 1,MA
         KA = (I – 1)*ZS + 1
         KB = KA + N*SS
         PROTO2(I) = N + I
912      READ 911,(A(K),K = KA,KB,SS)                                             35
911      FORMAT(10F8.3)
         DO 914 K = 1,N
914      PROTO1(K) = K
         FALL = 9
         CALL MATADR(A,N,M,PROTO1,PROTO2,ZS,SS)                                   40
         D = (N + 1)*(M + 2)
         IF(MET.EQ.1) GO TO 100
         IF(MET.EQ.2) GO TO 200
         IF(MET.EQ.3) GO TO 300
         IF(MET.EQ.5) GO TO 500                                                   45
         IF(MET.EQ.6) GO TO 600
100      PRINT 1010
         CALL SIMPLX(A,D,ZS,SS,N,M1,M2,M3,FALL,PROTO1,N,PROTO2,M,
       1    L1,N,L2,MK,L3,M2,EPS)
         GO TO 9000                                                              50
200      PRINT 1020
         CALL DUSEX(A,D,ZS,SS,N,M,FALL,.FALSE.,PROTO1,N,PROTO2,M,
       1    L1,N,L2,M,EPS)
         GO TO 9000
300      PRINT 1030                                                              55
         DB = MA*MA
         CALL RESEX(A,D,B,DB,C,MA,N,M,ZS,SS,FALL,PROTO1,N,PROTO2,M,
       1       L1,N,L2,M,L3,M,EPS)
         GO TO 9000
500      PRINT 1050                                                              60
         CALL DUOPLX(A,D,N,M1,M2,ZS,SS,FALL,PROTO1,N,PROTO2,M,
       1       L1,N,L2,M,L3,M2,EPS)
         GO TO 9000
```

```
 600   PRINT 1060
       CALL GOMORY(A,D,N,M,M1,M2,M3,ZS,SS,FALL,PROTO1,N,PROTO2,M,              65
      1     L1,N,L2,M,L3,M2,EPS)
1010   FORMAT(//* STANDARD SIMPLEX METHOD*)
1020   FORMAT(//* DUAL STANDARD SIMPLEX METHOD*)
1030   FORMAT(//* REVISED SIMPLEX METHOD*)
1050   FORMAT(//* STANDARD DUOPLEX METHOD*)                                    70
1060   FORMAT(//* INTEGER ALGORITHM OF GOMORY*)
9000   IF(FALL.EQ.0) PRINT 950
       IF(FALL.EQ.1) PRINT 951
       IF(FALL.EQ.2) PRINT 952
 950   FORMAT(//* FINITE SOLUTION*)                                            75
 951   FORMAT(//* SOLUTION UNBOUNDED*)
 952   FORMAT(//* NO FEASIBLE SOLUTION*)
       CALL MATADR(A,N,M,PROTO1,PROTO2,ZS,SS)
       EXMPL = EXMPL + 1
       GO TO 900                                                              80
 700   PRINT 930
 930   FORMAT(//* QUADRATIC OPTIMIZATION BY BEALE*)
       NA1 = N + 1
       L = 2*N + 1
       DO 931 I = 1,NA1                                                        85
       PROTO1(I) = I
       A(I) = 0.
       READ 932,(X(K),K = 1,NA1)
 932   FORMAT(10F8.3)
       DO 933 K = I,NA1                                                        90
       KH = (L − I + 1)*(I − 1)/2 + K
 933   C(KH) = X(K)
 931   CONTINUE
       CALL MATCDR(C,N,PROTO1)
       IF(M.EQ.0) GO TO 937                                                    95
       DO 936 I = 1,M
       PROTO2(I) = N + I
       KA1 = I*ZS + 1
       KB1 = KA1 + N*SS
 936   READ 911,(A(K),K = KA1,KB1,SS)                                         100
 937   CALL MATADR(A,N,M,PROTO1,PROTO2,ZS,SS)
       DC = ((N + 1)*(N + 2))/2
       DA = (M + N + 2)*(N + 1)
       CALL BEALE(C,DC,A,DA,N,M,ZS,SS,PROTO1,N,PROTO2,M + N,LIST,N,
      1     ABLIST,N,L1,N,L2,M + N,FALL,EPS)                                  105
       IF(FALL.EQ.0) PRINT 950
       IF(FALL.EQ.1) PRINT 951
       CALL MATCDR(C,N,PROTO1)
       CALL MATADR(A,N,M,PROTO1,PROTO2,ZS,SS)
       EXMPL = EXMPL + 1                                                      110
       GO TO 900
 800   PRINT 940
 940   FORMAT(//* QUADRATIC OPTIMIZATION BY WOLFE*)
       ZS = 3*N + M + 2
       SS = 1                                                                 115
       M2 = M + N + 1
       DA = ZS*M2
       DX = 4*N + 2*M + 1
       DO 941 I = 1,DA
 941   A(I) = 0.                                                              120
       MN = M + N
       DO 942 I = 1,MN
       PROTO2(I) = ZS − 1 + I
       IA = I*ZS + 1
       IE = IA + N*SS                                                         125
 942   READ 943,(A(K),K = IA,IE,SS)
```

```
943  FORMAT(10F8.3)
     NH = 3*N + M + 1
     DO 944 K = 1,NH
944  PROTO1(K) = K                                                    130
     CALL MATADR(A,N,M + N,PROTO1,PROTO2,ZS,SS)
     DO 945 I = 1,N
     M3 = (M + I)*ZS + 1
     M31 = M3 + (N + I)*SS
     M32 = M3 + (M + N + N + I)*SS                                     135
     A(M31) = A(M32) = 1.
     M33 = M3 + ZS - 1
     A(M33) = A(M3)
     A(M3) = 0.
     DO 945 K = 1,M                                                   140
     M34 = M3 + (N + N + K)*SS
     MHH = K*ZS + I*SS + 1
945  A(M34) = A(MHH)
     CALL MATADR(A,ZS - 1,M + N,PROTO1,PROTO2,ZS,SS)
     CALL WOLFE(A,DA,X,X0,DX,ZS,SS,N,M,PROTO1,NH,PROTO2,MN,FALL,       145
    1     L1,NH,L2,MN,LIST1,MN + 1,EPS)
     IF(FALL.EQ.1) PRINT 951
     IF(FALL.EQ.2) PRINT 952
     EXMPL = EXMPL + 1
     IF(FALL.GT.0) GO TO 900                                          150
     PRINT 950
     CALL MATADR(A,ZS - 1,M + N,PROTO1,PROTO2,ZS,SS)
     PRINT 960
960  FORMAT(//* X - VALUES OF THE ORIGINAL INDEPENDENT VARIABLES*//)
     PRINT 961,(X(I),I = 1,N)                                         155
961  FORMAT(20X,10F10.3)
     GO TO 900
999  CONTINUE
     END
C                                                                     160
C
C    SUBROUTINE MATADR
C
     SUBROUTINE MATADR (A,N,M,P1,P2,ZSCHR,SSCHR)
     REAL A(1)                                                        165
     INTEGER P1(1),P2(1),ZSCHR,SSCHR
     N1 = N + 1
     M1 = M + 1
     IF (N1.GT.11) N1 = 11
     N2 = N1 - 1                                                      170
     PRINT 101
101  FORMAT(//* LINEAR*//)
     PRINT 105,(K,K = 1,N2)
     PRINT 105,(P1(K),K = 1,N2)
     DO 1 I = 1,M1                                                    175
     K1 = (I - 1)*ZSCHR + 1
     K2 = K1 + N2*SSCHR
     IF(I.EQ.1)PRINT 102,(A(K),K = K1,K2,SSCHR)
102  FORMAT(/7X,F10.3,3X,10F10.3)
    1 IF(I.NE.1)PRINT 103,P2(I - 1),(A(K),K = K1,K2,SSCHR)            180
103  FORMAT(I5,2X,F10.3,3X,10F10.3)
    2 IF(N2.EQ.N)RETURN
     N1 = N2 + 1
     N2 = N
     IF(N2 - N1.GT.9)N2 = N1 + 9                                      185
     PRINT 104,(K,K = N1,N2)
104  FORMAT(//18X,10I10)
     PRINT 105,(P1(K),K = N1,N2)
```

```
      105   FORMAT(18X,10I10)
            DO 3 I = 1,M1
            K1 = (I − 1)∗ZSCHR + N1∗SSCHR + 1
            K2 = K1 + (N2 − N1)∗SSCHR
            IF(I.EQ.1)PRINT 106,(A(K),K = K1,K2,SSCHR)
      106   FORMAT(/20X,10F10.3)
        3   IF(I.NE.1)PRINT 107,P2(I − 1),(A(K),K = K1,K2,SSCHR)
      107   FORMAT(I5,15X,10F10.3)
            GO TO 2
            END
C
C
C     SUBROUTINE MATBDR
C
            SUBROUTINE MATBDR(B,M,ZSCHR,SSCHR)
            REAL B(1)
            INTEGER ZSCHR,SSCHR
            PRINT 900
      900   FORMAT(//∗ MATRIX B∗)
            M1 = 1
            M2 = M
            MH = M + 1
        2   IF(M2 − M1.GT.9) M2 = M1 + 9
            PRINT 910
      910   FORMAT(//)
            DO 1 I = 1,MH
            K1 = (I − 1)∗ZSCHR + M1∗SSCHR + 1
            K2 = K1 + (M2 − M1)∗SSCHR
        1   PRINT 100,(B(K),K = K1,K2,SSCHR)
      100   FORMAT(20X,10F10.3)
            IF(M2.EQ.M) RETURN
            M1 = M2 + 1
            M2 = M
            GO TO 2
            END
C
C
C     SUBROUTINE MATCDR
C
            SUBROUTINE MATCDR (C,N,P1)
            INTEGER P1(1)
            REAL C(1),X(50)
            L = 2∗N + 1
            N1 = NN = N + 1
            IF (N1.GT.11) N1 = 11
            N2 = N1 − 1
            PRINT 101
      101   FORMAT(//∗ QUADRATIC OBJECTIVE FUNCTION∗)
            PRINT 105,(K,K = 1,N2)
            PRINT 106,(P1(K),K = 1,N2)
      105   FORMAT(//18X,10I10)
      106   FORMAT(18X,10I10)
            DO 1 I = 1,NN
            DO 2 K = 1,N1
            IF(K.LT.I) 10,11
       10   K1 = (L − K + 1)∗(K − 1)/2 + I
            GO TO 2
       11   K1 = (L − I + 1)∗(I − 1)/2 + K
        2   X(K) = C(K1)
            IF(I.EQ.1) PRINT 103,(X(K),K = 1,N1)
      103   FORMAT(/5X,F12.3,3X,10F10.3)
        1   IF(I.NE.1) PRINT 104,P1(I − 1),(X(K),K = 1,N1)
      104   FORMAT(X,I4,F12.3,3X,10F10.3)
```

190

195

200

205

210

215

220

225

230

235

240

245

250

```
     9  IF(N2.EQ.N) RETURN
        N1 = N2 + 1
        N2 = N
        IF(N2 − N1.GT.9) N2 = N1 + 9                                    255
        PRINT 105,(K,K = N1,N2)
        PRINT 106,(P1(K),K = N1,N2)
        DO 3 I = 1,NN
        DO 4 K = N1,N2
        KH = K + 1                                                      260
        IF(KH.LT.I) 14,15
    14  K1 = (L − KH + 1)∗(KH − 1)/2 + I
        GO TO 4
    15  K1 = (L − I + 1)∗(I − 1)/2 + KH
     4  X(K) = C(K1)                                                    265
        IF(I.EQ.1) PRINT 107,(X(K),K = N1,N2)
   107  FORMAT(/20X,10F10.3)
     3  IF(I.NE.1) PRINT 108,P1(I − 1),(X(K),K = N1,N2)
   108  FORMAT(X,I4,15X,10F10.3)
        GO TO 9                                                         270
        END
C
C
C       SUBROUTINE MP1
C                                                                       275
C       MP1 ORDERS THE INTEGERS IN THE LIST L1(K) SUCH THAT THE
C       CORRESPONDING REAL NUMBERS A(ZNR∗ZSCHR + L1(K)∗SSCHR) DECREASE
C       MONOTONICALLY WITH K.
C
        SUBROUTINE MP1(A,J1,ZNR,ZSCHR,SSCHR,L1,L10,JL1)                 280
        INTEGER R,S,ZSCHR,SSCHR,ZNR
        DIMENSION A(J1),L1(JL1)
        IF(L10.LT.2)RETURN
        DO 1 K = 2,L10
        I = 0                                                           285
        KK = K − 1
        DO 2 S = 1,KK
        KH = ZNR∗ZSCHR + 1
        KH0 = KH + L1(K)∗SSCHR
        KH1 = K − S                                                     290
        KH = KH + L1(KH1)∗SSCHR
     2  IF(A(KH0).GT.A(KH)) I = I + 1
        IF(I.EQ.0) GO TO 1
        R = L1(K)
        DO 3 S = 1,I                                                    295
        KK1 = K − S
     3  L1(KK1 + 1) = L1(KK1)
        KK1 = K − I
        L1(KK1) = R
     1  CONTINUE                                                        300
        RETURN
        END
C
C
C       SUBROUTINE MP2                                                  305
C
C       MP2 DETERMINES THE MINIMUM OF ALL THOSE V∗A(I∗ZSCHR)/A(I∗ZSCHR + KP
C       ∗SSCHR) FOR WHICH THE A(I∗ZSCHR) ARE NONNEGATIVE AND THE V∗A(I∗
C       ZSCHR + KP∗SSCHR) ARE POSITIVE.  DEGENERACY IS TAKEN INTO CONSIDER −
C       TION.                                                           310
C
        SUBROUTINE MP2(A,J1,L2,L20,JL2,IP,ZSCHR,SSCHR,KP,Q1,N,IV,EPS)
        DIMENSION A(J1),L2(JL2)
        INTEGER ZSCHR,SSCHR,Z
```

```
                                                                        315
      V = IV
      IP = 0
      IF(L20.LT.1) RETURN
      DO 1 I = 1,L20
      KH = L2(I)*ZSCHR + 1                                              320
      KH1 = KH + KP*SSCHR
      IF(V*A(KH1).GT.EPS) GO TO 2
    1 CONTINUE
      RETURN
    2 Q1 = V*A(KH)/A(KH1)                                               325
      IP = L2(I)
      Z = I + 1
      IF(Z.GT.L20) RETURN
      DO 3 I = Z,L20
      KH = L2(I)*ZSCHR + 1                                              330
      KH1 = KH + KP*SSCHR
      IF(V*A(KH1).LE.EPS) GO TO 3
      Q = V*A(KH)/A(KH1)
      IF(Q.GE.Q1) GO TO 4
      IP = L2(I)                                                        335
      Q1 = Q
      GO TO 3
    4 IF(Q.NE.Q1) GO TO 3
C
C     HERE IT IS DETERMINED WHICH OF TWO ROWS WITH THE SAME QUOTIENT
C     QUALIFIES AS PIVOT ROW.                                          340
C
      I0 = L2(I)
      DO 5 K = 1,N
      KH0 = IP*ZSCHR + K*SSCHR + 1                                      345
      KH2 = IP*ZSCHR + KP*SSCHR + 1
      KH = I0*ZSCHR + K*SSCHR + 1
      QP = V*A(KH0)/A(KH2)
      Q0 = V*A(KH)/A(KH1)
      IF(QP.LT.Q0) GO TO 3
      IF(Q0.LT.QP) GO TO 6                                             350
    5 CONTINUE
    6 IP = I0
    3 CONTINUE
      RETURN
      END                                                              355
C
C
C
C     SUBROUTINE MP3
C
C     MP3 EXCHANGES A BASIC AND A NONBASIC VARIABLE.  I0 AND I1 SPECIFY  360
C     THE ROWS AND K0 AND K1 THE COLUMNS TO WHICH THE TRANSFORMATION
C     IS TO APPLY.  P1 AND P2 INDICATE WHETHER THE PIVOT ROW AND PIVOT
C     COLUMN, RESPECTIVELY, HAVE ALREADY BEEN TRANSFORMED.
C
      SUBROUTINE MP3(A,J1,I0,I1,K0,K1,IP,KP,ZSCHR,SSCHR,P1,P2)         365
      DIMENSION A(J1)
      INTEGER ZSCHR,SSCHR,P1,P2
      KH = IP*ZSCHR + KP*SSCHR + 1
      PIV = 1./A(KH)
      II0 = I0 + 1                                                     370
      II1 = I1 + 1
      KK0 = K0 + 1
      KK1 = K1 + 1
      IF(II0.GT.II1) GO TO 6
      DO 1 II = II0,II1                                                375
      I = II - 1
      IF(I.EQ.IP) GO TO 1
```

```
       KH0 = I*ZSCHR + KP*SSCHR + 1
       IF(P2.EQ.1)A(KH0) = A(KH0)*PIV
       DO 2 KK = KK0,KK1                                        380
       K = KK - 1
       IF(K.EQ.KP) GO TO 2
       KH1 = I*ZSCHR + K*SSCHR + 1
       KH2 = IP*ZSCHR + K*SSCHR + 1
       A(KH1) = A(KH1) - A(KH2)*A(KH0)                          385
    2  CONTINUE
    1  CONTINUE
    6  IF(P1.NE.1) GO TO 4
       DO 5 KK = KK0,KK1
       K = KK - 1                                               390
       KH2 = IP*ZSCHR + K*SSCHR + 1
    5  IF(K.NE.KP)A(KH2) = - A(KH2)*PIV
    4  IF(P2.EQ.1)A(KH) = PIV
       RETURN
       END                                                      395
C
C
C      SUBROUTINE MP5
C
C      MP5 DETERMINES THE MINIMUM OF THE NUMBERS A(ZNR*ZSCHR + L1(K)*   400
C      SSCHR) FOR K = 1,2,....,L10, KP IS THE INDEX K FOR WHICH THE
C      MINIMUM Q1 IS ASSUMED.
C
       SUBROUTINE MP5(A,J1,L1,L10,JL1,Q1,KP,ZNR,ZSCHR,SSCHR)
       INTEGER ZNR,ZSCHR,SSCHR                                  405
       DIMENSION A(J1),L1(JL1)
       KP = L1(1)
       KH = ZNR*ZSCHR + 1
       KH0 = KH + L1(1)*SSCHR
       Q1 = A(KH0)                                              410
       IF(L10.LT.2) RETURN
       DO 1 K = 2,L10
       KH0 = KH + L1(K)*SSCHR
       IF(A(KH0).GE.Q1) GO TO 1
       Q1 = A(KH0)                                              415
       KP = L1(K)
    1  CONTINUE
       RETURN
       END                                                      420
C
C
C      SUBROUTINE MP7
C
C      MP7 DETERMINES THE MAXIMUM OF THE ELEMENTS
C      A(ZNR*ZSCHR + L1(K)*SSCHR)                               425
C
       SUBROUTINE MP7(A,J1,ZNR,ZSCHR,SSCHR,KP,L1,L10,JL1,MAX)
       DIMENSION A(J1),L1(JL1)
       INTEGER ZNR,ZSCHR,SSCHR
       REAL MAX                                                 430
       KH = ZNR*ZSCHR + L1(1)*SSCHR + 1
       MAX = A(KH)
       KP = L1(1)
       IF(L10.LT.2) RETURN
       DO 1 K = 2,L10                                           435
       KH = ZNR*ZSCHR + L1(K)*SSCHR + 1
       IF(A(KH).LE.MAX) GO TO 1
       MAX = A(KH)
       KP = L1(K)
```

```
    1   CONTINUE                                                      440
        RETURN
        END
C
C
C       SUBROUTINE MP8                                                445
C
C       DETERMINATION OF THE MAXIMUM OF THE NUMBERS ABS(AZNR•
C       ZSCHR + LIST(K)•SSCHR)) FOR K = 1,2,....,LIST0
C
        SUBROUTINE MP8(A,J1,ZNR,ZSCHR,SSCHR,LIST,LIST0,JLIST,KP,MAX)   450
        INTEGER ZNR,ZSCHR,SSCHR
        REAL MAX
        DIMENSION A(J1),LIST(JLIST)
        KP = LIST(1)
        KH = ZNR•ZSCHR + 1                                            455
        KH0 = LIST(1)•SSCHR + KH
        MAX = A(KH0)
        IF(LIST0.LT.2) RETURN
        DO 1 K = 2,LIST0
        KH0 = KH + LIST(K)•SSCHR                                      460
        IF(ABS(MAX).GT.ABS(A(KH0))) GO TO 1
        KP = LIST(K)
        MAX = A(KH0)
    1   CONTINUE
        RETURN                                                        465
        END
C
C
C       SUBROUTINE MP9
C                                                                     470
C       ROUTINE FOR THE TRANSFORMATION OF THE QUADRATIC FORM IN
C       BEALE'S METHOD
C       COMMENTS REFER TO FIGURE 16 IN SECTION 3.5.7. OF THE BOOK
C
        SUBROUTINE MP9(A,J1,C,JC,IP,KP,N,M1,ZSCHR,SSCHR,B1)           475
        INTEGER ZSCHR,SSCHR,R,S,T,Z,Z1
        LOGICAL B1
        DIMENSION A(J1),C(JC)
        IN = N + 1
        DO 9000 IR = 1,IN                                             480
        R = IR - 1
        IF(R.EQ.KP) GO TO 9000
        Z = 0
        IF(R.GT.KP)Z1 = KP - 1
        IF(R.LT.KP)Z1 = R                                             485
        IZ1 = Z1 + 1
        DO 9002 IS = 1,IZ1
        S = IS - 1
        KH = R + Z + 1
        KH1 = IP•ZSCHR + R•SSCHR + 1                                  490
        KH0 = KP + Z + 1
        C(KH) = C(KH) + A(KH1)•C(KH0)
   9002 Z = Z + N - S
        T = Z + KP
   9000 CONTINUE                                                      495
C
C       THE BLOCK UP TO 9000 TRANSFORMS THE
C       HATCHED ELEMENTS
C
   9100 IKP = KP + 1                                                  500
        IF(IKP.GT.N) GO TO 9300
        DO 9102 S = IKP,N
```

```
       DO 9103 R = S,N
       KH = Z + N − KP + R + 1
       KH1 = IP*ZSCHR + R*SSCHR + 1                                505
       KH0 = T + S − KP + 1
 9103  C(KH) = C(KH) + A(KH1)*C(KH0)
 9102  Z = Z + N − S
C
C      THE BLOCK 91XX TRANSFORMS THE NONHATCHED PART               510
C      OF THE ELEMENTS
C
 9200  DO 9201 R = IKP,N
       KH = T + R − KP + 1
       KH1 = IP*ZSCHR + R*SSCHR + 1                                515
 9201  C(KH) = C(KH) + A(KH1)*C(T + 1)
C
C      THE BLOCK 92XX TRANSFORMS THE ELEMENTS
C      DESIGNATED BY 1
C                                                                  520
 9300  Z1 = 0
       DO 9301 IR = 1,KP
       R = IR − 1
       Z = 0
       KH = KP + Z1 + 1                                            525
       KH1 = IP*ZSCHR + R*SSCHR + 1
       STORE = C(KH) + A(KH1)*C(T + 1)
       DO 9302 IS = 1,IR
       S = IS − 1
       KH = IP*ZSCHR + S*SSCHR + 1                                 530
       KH1 = IR + Z
       C(KH1) = C(KH1) + A(KH)*STORE
 9302  Z = Z + N − S
 9301  Z1 = Z1 + N − R
C                                                                  535
C      THE BLOCK 93XX TRANSFORMS FOR A SECOND TIME
C      THE DIAGONALLY HATCHED ELEMENTS CONTAINED
C      IN THE UPPER TRIANGLE
C
 9400  IF(IKP.GT.N) GO TO 9500                                     540
       DO 9401 R = IKP,N
       Z = 0
       IR = R + 1
       DO 9401 IS = 1,IR
       S = IS − 1                                                  545
       IF(S.NE.KP) GO TO 9402
       Z = Z + 1
       GO TO 9401
 9402  KH = IR + Z
       KH1 = IP*ZSCHR + S*SSCHR + 1                                550
       KH0 = T + R − KP + 1
       C(KH) = C(KH) + A(KH1)*C(KH0)
 9401  Z = Z + N − S
C
 9500  IF(.NOT.B1) GO TO 9600                                      555
       Z = 0
       KH0 = IP*ZSCHR + KP*SSCHR + 1
       DO 9501 IS = 1,KP
       S = IS − 1
       KH = IKP + Z                                                560
       KH1 = IP*ZSCHR + S*SSCHR + 1
       C(KH) = C(KH) + A(KH1)*C(T + 1)
 9501  Z = Z + N − S
       DO 9503 R = KP,N
       KH = R + T − KP + 1                                         565
```

```
 9503   C(KH) = C(KH)•A(KHO)
        Z = 0
        DO 9504 IS = 1,IKP
        S = IS − 1
        KH = IKP + Z                                                    570
        C(KH) = C(KH)•A(KHO)
 9504   Z = Z + N − S
        RETURN
C
C       THE BLOCKS 94XX AND 95XX TRANSFORM THE ELEMENTS IN THE HATCHED  575
C       RECTANGLE AND THE ELEMENTS IN THE NONHATCHED TRIANGLE (UP TO
C       9401), AS WELL AS THE ELEMENTS DESIGNATED BY 1 AND 2, PROVIDED
C       THE PIVOT ROW IS CONTAINED IN THE CONSTRAINT SECTION.
C
 9600   Z = 0                                                           580
        DO 9601 IS = 1,KP
        S = IS − 1
        KH = KP + Z + 1
        C(KH) = 0.
 9601   Z = Z + N − S                                                   585
        C(T + 1) = 1./C(T + 1)
        IF(IKP.GT.N) GO TO 9602
        DO 9603 R = IKP,N
        KH = R + T − KP + 1
 9603   C(KH) = 0.                                                      590
 9602   RETURN
        END
C
C       SUBROUTINE MP10                                                 595
C
C       MP10 DESCRIBES A PROCEDURE FOR THE STEPWISE
C       COMPUTATION OF THE INVERSE MATRIX NEEDED
C       IN THE REVISED SIMPLEX METHOD
C                                                                       600
        SUBROUTINE MP10(B,JB,C,JC,M,IP,S,ZSCHR1,SSCHR1)
        INTEGER ZSCHR1,SSCHR1,S
        DIMENSION B(JB),C(JC)
    1   DO 2 K = 1,S
        KH = IP•ZSCHR1 + K•SSCHR1 + 1                                   605
        IF(B(KH).EQ.0.) GO TO 2
        IM = M + 1
        DO 3 II = 1,IM
        I = II − 1
        IF(I.EQ.IP) GO TO 3                                             610
        IF(C(II).EQ.0.) GO TO 3
        KH = K•SSCHR1 + 1
        KHO = KH + I•ZSCHR1
        KH1 = KH + IP•ZSCHR1
        B(KHO) = B(KHO) + B(KH1)•C(II)                                  615
    3   CONTINUE
    2   CONTINUE
        DO 4 K = 1,S
        KH = IP•ZSCHR1 + K•SSCHR1 + 1
    4   B(KH) = B(KH)•C(IP + 1)                                         620
        RETURN
        END
C
C
C       ROUTINE FOR THE OPTIMIZATION OF A LINEAR PROGRAM                625
C       USING THE SIMPLEX METHOD.
C       DEGENERACIES ARE TAKEN INTO ACCOUNT
C       SIMPLEX USES THE SUBROUTINES MP2,MP3,MP7 AND MP8
C
```

```
      SUBROUTINE SIMPLX (A,J1,ZSCHR,SSCHR,N,M1,M2,M3,FALL,          630
     1     PROTO1,IPROT1,PROTO2,IPROT2,L1,JL1,L2,JL2,L3,JL3,EPS)
      DIMENSION A(J1),PROTO1(IPROT1),PROTO2(IPROT2),
     1     L1(JL1),L2(JL2),L3(JL3)
      INTEGER ZSCHR,SSCHR,V,R,S,PROTO1,PROTO2,FALL
      REAL MAX                                                       635
C
C     INITIALIZATION OF THE INDEX LISTS
C
      M123 = M1 + M2 + M3
      R = 0                                                          640
      V = - 1
      DO 1 K = 1,N
      L1(K) = K
    1 PROTO1(K) = K
      L10 = N                                                        645
      DO 2 I = 1,M123
    2 L2(I) = I
      L20 = M123
      DO 3 I = 1,M123
    3 PROTO2(I) = N + I                                              650
      IF(M2 + M3.EQ.0) GO TO 103
C
C     IF THE ORIGIN IS A FEASIBLE SOLUTION THE BLOCK
C     UP TO 103 CAN BE BYPASSED
C                                                                    655
      IF(M2.EQ.0) GO TO 22
      DO 17 I = 1,M2
   17 L3(I) = 1
C
C     COMPUTATION OF THE AUXILIARY OBJECTIVE FUNCTION                660
C     IN PREPARATION FOR THE M — METHOD
C
   22 R = 1
      N1 = N + 1
      DO 4 K = 1,N1                                                  665
      KK = K - 1
      Q1 = 0.
      N2 = M1 + 1
      DO 5 I = N2,M123
      KH = I*ZSCHR + KK*SSCHR + 1                                    670
    5 Q1 = Q1 + A(KH)
      KH = (M123 + 1)*ZSCHR + KK*SSCHR + 1
    4 A(KH) = - Q1
C
C     COMPUTATION OF A FEASIBLE SOLUTION BY THE SIMPLEX METHOD       675
C     USING THE ABOVE — CALCULATED AUXILIARY OBJECTIVE FUNCTION.
C     MP7 DETERMINES THE MAXIMAL COEFFICIENT OF THE AUXILIARY
C     OBJECTIVE FUNCTION.
C
  100 CALL MP7(A,J1,M123 + 1,ZSCHR,SSCHR,KP,L1,L10,JL1,MAX)          680
      KH = (M123 + 1)*ZSCHR + 1
      IF(MAX.GT.EPS.OR.A(KH).GE. - EPS) GO TO 6
      FALL = 2
      RETURN
C                                                                    685
C     IF THE MAXIMAL COEFFICIENT OF THE AUXILIARY OBJECTIVE FUNCTION
C     IS NONPOSITIVE AND THE VALUE OF THE AUXILIARY OBJECTIVE FUNCTION
C     IS NEGATIVE, NO FEASIBLE SOLUTION EXISTS
C
    6 IF(MAX.GT.EPS.OR.A(KH).GT.EPS) GO TO 16                        690
      M12 = M1 + M2 + 1
      IF(M12.GT.M123) GO TO 30
```

```
        DO 18 IP = M12,M123
        IF(PROTO2(IP).NE.IP + N) GO TO 18
        CALL MP8(A,J1,IP,ZSCHR,SSCHR,L1,L10,JL1,KP,MAX)                      695
        IF(MAX.GT.0.) GO TO 7
    18  CONTINUE
    30  R = 0
        M12 = M12 − 1
        IF(N2.GT.M12) GO TO 103                                              700
        DO 20 I = N2,M12
        KH = I − M1
        IF(L3(KH).NE.1) GO TO 20
        DO 21 K = 1,N1
        KH = I*ZSCHR + (K − 1)*SSCHR + 1                                     705
    21  A(KH) = − A(KH)
    20  CONTINUE
        GO TO 103
C
C       IF THE MAXIMAL COEFFICIENT OF THE AUXILIARY OBJECTIVE FUNCTION       710
C       IS NONPOSITIVE AND THE VALUE OF THIS FUNCTION EQUALS ZERO,
C       THEN A FEASIBLE SOLUTION HAS BEEN FOUND AND CONTROL CAN BE
C       TRANSFERRED TO THE OPTIMIZATION SECTION.
C
    16  CALL MP2(A,J1,L2,L20,JL2,IP,ZSCHR,SSCHR,KP,Q1,N,V,EPS)               715
C
C       MP2 ENSURES THAT NO CONSTRAINT IS VIOLATED IN THE EXCHANGE
C
        IF(IP.NE.0) GO TO 7
        FALL = 2                                                             720
        RETURN
C
C       SINCE THE MAXIMUM OF THE AUXILIARY OBJECTIVE FUNCTION IS
C       PLUS INFINITY, THERE EXISTS NO FEASIBLE SOLUTION.
C                                                                            725
     7  CALL MP3(A,J1,0,M123 + 1,0,N,IP,KP,ZSCHR,SSCHR,1,1)
        IF(PROTO2(IP).LT.N + M1 + M2 + 1) GO TO 101
C
C       IF THE VARIABLE JUST REMOVED FROM THE BASIS BY MP3 CORRESPONDS
C       TO AN EQUALITY CONSTRAINT, THE NEXT BLOCK ENSURES THAT THIS          730
C       VARIABLE REMAINS OUTSIDE THE BASIS.  OTHERWISE THE BLOCK
C       UP TO 101 MUST BE BYPASSED.
C
        DO 8 K = 1,L10
        IF(L1(K).EQ.KP) GO TO 9                                              735
     8  CONTINUE
     9  L10 = L10 − 1
        DO 10 S = K,L10
    10  L1(S) = L1(S + 1)
        GO TO 102                                                            740
C
   101  IF(PROTO2(IP).LT.N + M1 + 1) GO TO 1021
        KH = PROTO2(IP) − M1 − N
        IF(L3(KH).EQ.0) GO TO 1021
        L3(KH) = 0                                                           745
   102  KH = (M123 + 1)*ZSCHR + KP*SSCHR + 1
        A(KH) = A(KH) + 1.
        MH = M123 + 2
        DO 12 I = 1,MH
        KH = (I − 1)*ZSCHR + KP*SSCHR + 1                                    750
    12  A(KH) = − A(KH)
C
C       IN THE FOLLOWING BLOCK UP TO 103 THE PROTO LISTS
C       ARE UPDATED
C                                                                            755
```

```
  1021   S = PROTO1(KP)
         PROTO1(KP) = PROTO2(IP)
         PROTO2(IP) = S
         CALL MATADR(A,N,M123,PROTO1,PROTO2,ZSCHR,SSCHR)
         IF(R.NE.0) GO TO 100                                    760
C
C        OPTIMIZATION SECTION
C
   103   CALL MP7(A,J1,0,ZSCHR,SSCHR,KP,L1,L10,JL1,MAX)
         IF(MAX.GT.0.) GO TO 14                                  765
         FALL = 0
         RETURN
C
C        IF THE MAXIMAL COEFFICIENT OF THE OBJECTIVE FUNCTION IS LESS THAN
C        OR EQUAL TO ZERO, THE OPTIMAL TABLEAU HAS BEEN OBTAINED  770
C
    14   CALL MP2(A,J1,L2,L20,JL2,IP,ZSCHR,SSCHR,KP,Q1,N,V,EPS)
   104   IF(IP.NE.0) GO TO 15
         FALL = 1
         RETURN                                                  775
    15   CALL MP3(A,J1,0,M123,0,N,IP,KP,ZSCHR,SSCHR,1,1)
         GO TO 1021
         END
C
C                                                                780
C        ROUTINE FOR DETERMINING THE MINIMUM OF A LINEAR PROGRAM BY MEANS
C        OF THE DUAL SIMPLEX METHOD.  THE TABLEAU IS ASSUMED TO BE DUALLY
C        FEASIBLE.  DUSEX USES THE SUBROUTINES MP2,MP3,MP5
C
         SUBROUTINE DUSEX(A,J1,ZSCHR,SSCHR,N,M,FALL,W,PROTO1,     785
     1        IPROT1,PROTO2,IPROT2,L1,JL1,L2,JL2,EPS)
         DIMENSION A(J1),L1(JL1),L2(JL2)
         INTEGER PROTO1(IPROT1),PROTO2(IPROT2),FALL,ZSCHR,SSCHR,V,ZNR
         LOGICAL W
         DO 1 K = 1,M                                            790
     1   L1(K) = K
         L10 = M
         DO 2 I = 1,N
     2   L2(I) = I
         L20 = N                                                 795
         V = 1
         IF(W) GO TO 100
         DO 3 K = 1,N
     3   PROTO1(K) = K
         DO 4 I = 1,M                                            800
     4   PROTO2(I) = N + I
C
   100   I = ZSCHR
         ZSCHR = SSCHR
         SSCHR = I
         I = M                                                   805
         M = N
         N = I
         CALL MP5(A,J1,L1,L10,JL1,Q1,KP,0,ZSCHR,SSCHR)
         IF(Q1.LT.0.) GO TO 101                                  810
         FALL = 0
         GO TO 300
   101   CALL MP2(A,J1,L2,L20,JL2,IP,ZSCHR,SSCHR,KP,Q1,N,V,EPS)
         IF(IP.NE.0) GO TO 200
         FALL = 2
         GO TO 300                                               815
C
   200   I = ZSCHR
```

```
              ZSCHR = SSCHR
              SSCHR = I
              I = M                                                          820
              M = N
              N = I
              I = KP
              KP = IP
              IP = I                                                         825
              I = PROTO1(KP)
              PROTO1(KP) = PROTO2(IP)
              PROTO2(IP) = I
              CALL MP3(A,J1,0,M,0,N,IP,KP,ZSCHR,SSCHR,1,1)
              CALL MATADR(A,N,M,PROTO1,PROTO2,ZSCHR,SSCHR)                   830
              GO TO 100
    C
       300    I = ZSCHR
              ZSCHR = SSCHR
              SSCHR = I                                                      835
              I = M
              M = N
              N = I
              RETURN
              END                                                           840
    C
    C
    C         ROUTINE FOR DETERMINING THE MAXIMUM OF A LINEAR PROGRAM BY THE
    C         REVISED SIMPLEX METHOD.  RESEX PRESUPPOSES THAT THE ORIGIN IS
    C         A FEASIBLE SOLUTION.  RESEX USES THE SUBROUTINES MP7 AND       845
    C         MP10.
    C
              SUBROUTINE RESEX(A,J1,B,JB,C,JC,N,M,ZSCHR,SSCHR,FALL,PROTO1,
            1         IPROT1,PROTO2,IPROT2,L1,JL1,L2,JL2,L3,JL3,EPS)
              DIMENSION A(J1),B(JB),C(JC),L1(JL1),L2(JL2),L3(JL3)            850
              INTEGER PROTO1(IPROT1),PROTO2(IPROT2),FALL,ZSCHR,SSCHR,S,
            1         ZSCHR1,SSCHR1
              REAL MAX
              DO 1 K = 1,N
              L1(K) = K                                                      855
         1    PROTO1(K) = K
              DO 2 I = 1,M
         2    PROTO2(I) = N + I
              L10 = N
              L20 = 0                                                        860
              ZSCHR1 = 1
              SSCHR1 = M + 1
    C
    C         GENERATION OF THE IDENTITY MATRIX
    C                                                                        865
              M1 = M + 1
              DO 3 I = 1,M1
              DO 3 K = 1,M
              KH = (I - 1)*ZSCHR1 + K*SSCHR1 + 1
              B(KH) = 0.                                                     870
         3    IF(K.EQ.I - 1) B(KH) = - 1.
    C
    C         DETERMINATION OF THE PIVOT COLUMN
    C                                                                        875
       100    MAX = 0.
              Q1 = 0.
              IF(L10.EQ.0) GO TO 200
              CALL MP7(A,J1,0,ZSCHR,SSCHR,KP,L1,L10,JL1,MAX)
       200    IF(L20.EQ.0) GO TO 300                                        880
              DO 201 I = 1,L20
```

```
        KH = L2(I)*SSCHR1 + 1
        IF(B(KH).LE.Q1) GO TO 201
        S = I
        Q1 = B(KH)                                                      885
201     CONTINUE
300     IF(MAX.GT.0..OR.Q1.GT.0.) GO TO 301
        FALL = 0
        RETURN
301     K = 0                                                           890
        IF(MAX.GE.Q1) GO TO 400
        K = 1
        KP = L3(S)
        DO 302 II = 1,M1
        I = II − 1                                                      895
        KH = I*ZSCHR1 + L2(S)*SSCHR1 + 1
302     C(II) = − B(KH)
        GO TO 500
400     KH = KP*SSCHR + 1
        C(1) = − A(KH)                                                  900
        DO 401 I = 1,M
        Q1 = 0.
        DO 402 J = 1,M
        KH1 = KH + J*ZSCHR
        KH2 = I*ZSCHR1 + J*SSCHR1 + 1                                   905
402     IF(A(KH1).NE.0..AND.B(KH2).NE.0.)Q1 = Q1 + A(KH1)*B(KH2)
401     C(I + 1) = Q1
C
C       DETERMINATION OF THE PIVOT ROW
C                                                                       910
500     IP = 0
        DO 501 I = 1,M
        IF(C(I + 1).GT.0.) GO TO 502
501     CONTINUE
        FALL = 1                                                        915
        RETURN
502     KH = I*ZSCHR + 1
        Q1 = A(KH)/C(I + 1)
        IP = I
600     J = IP                                                          920
        DO 601 I = J,M
        IF(C(I + 1).LE.0.) GO TO 601
        KH = I*ZSCHR + 1
        IF(A(KH)/C(I + 1).GE.Q1) GO TO 601
        Q1 = A(KH)/C(I + 1)                                             925
        IP = I
601     CONTINUE
        DO 602 II = 1,M1
602     IF(II − 1.NE.IP)C(II) = − C(II)/C(IP + 1)
        C(IP + 1) = 1./C(IP + 1)                                        930
C
C       TRANSFORMATION OF THE FIRST COLUMN OF TABLEAU A
C
        CALL MP10(A,J1,C,JC,M,IP,1,ZSCHR,0)
C                                                                       935
C       TRANSFORMATION OF THE FIRST ROW OF TABLEAU A
C
        DO 603 J = 1,N
        IF(J.EQ.KP) GO TO 603
        Q1 = 0.                                                         940
        DO 604 I = 1,M
        KH = IP*ZSCHR1 + I*SSCHR1 + 1
        KH1 = I*ZSCHR + J*SSCHR + 1
```

```
604   IF(B(KH).NE.0.)Q1 = Q1 + B(KH)*A(KH1)
      KH = J*SSCHR + 1                                              945
      A(KH) = A(KH) − Q1*C(1)
603   CONTINUE
      KH = KP*SSCHR + 1
      A(KH) = − C(1)
C                                                                   950
C     TRANSFORMATION OF THE INVERSE
C
610   CALL MP10(B,JB,C,JC,M,IP,M,ZSCHR1,SSCHR1)
      IF(K.NE.0) GO TO 701
      IF(L20.LT.1) GO TO 605                                        955
      DO 606 J = 1,L20
      IF(IP.EQ.L2(J)) GO TO 607
606   CONTINUE
      GO TO 605
607   IF(L3(J).EQ.KP) GO TO 800                                     960
      IF(L10.LT.1) GO TO 609
      DO 608 I = 1,L10
608   IF(L1(I).EQ.KP)L1(I) = L3(J)
609   L3(J) = KP
      GO TO 800                                                     965
605   L20 = L20 + 1
      L2(L20) = IP
      L3(L20) = KP
700   IF(L10.LT.1) GO TO 701
      DO 702 I = 1,L10                                              970
      IF(L1(I).EQ.KP) GO TO 704
702   CONTINUE
      GO TO 701
704   L10 = L10 − 1
      DO 703 J = I,L10                                              975
703   L1(J) = L1(J + 1)
      GO TO 800
701   IF(L20.LT.1) GO TO 707
      DO 705 J = 1,L20
      IF(IP.EQ.L2(J)) GO TO 800                                     980
705   CONTINUE
707   L10 = L10 + 1
      L1(L10) = L3(S)
      L20 = L20 − 1
      DO 706 J = S,L20                                              985
      L2(J) = L2(J + 1)
706   L3(J) = L3(J + 1)
800   K = PROTO1(KP)
      PROTO1(KP) = PROTO2(IP)
      PROTO2(IP) = K                                                990
      CALL MATADR(A,N,M,PROTO1,PROTO2,ZSCHR,SSCHR)
      CALL MATBDR(B,M,ZSCHR1,SSCHR1)
      GO TO 100
      END
C                                                                   995
C
C     ROUTINE FOR DETERMINING THE MAXIMUM OF A LINEAR PROGRAM.
C     DUOPLEX USES THE SUBROUTINES MP1,MP2,MP3, AND MP5.
C
      SUBROUTINE DUOPLX(A,J1,N,M1,M2,ZSCHR,SSCHR,FALL,PROTO1,        1000
     1   IPROT1,PROTO2,IPROT2,L1,JL1,L2,JL2,L3,JL3,EPS)
      DIMENSION A(J1),PROTO1(IPROT1),PROTO2(IPROT2),L1(JL1),L2(JL2),
     1   L3(JL3)
      INTEGER PROTO1,PROTO2,ZSCHR,SSCHR,FALL,R,S,Z
C                                                                   1005
C     IN THE FOLLOWING BLOCK THE INDEX LISTS
```

```
C       ARE INITIALIZED
C
        L10 = N
        J = 1                                                          1010
        R = 0
        DO 1 K = 1,N
        L1(K) = K
     1  PROTO1(K) = K
        M12 = M1 + M2                                                  1015
        DO 2 I = 1,M12
     2  PROTO2(I) = N + I
        IF(M2.NE.0) GO TO 100
        L30 = 0
        GO TO 300                                                      1020
C
C       IF THERE ARE NO EQUALITY CONSTRAINTS, THE
C       FOLLOWING BLOCK UP TO STATEMENT 300 CAN BE BYPASSED
C
   100  L30 = M2                                                       1025
        DO 101 I = 1,M2
   101  L3(I) = M1 + I
C
C       IN THE BLOCK WITH THE LABELS 1XX THE BASIC VARIABLES CORRESPONDING
C       TO THE EQUATIONS ARE REMOVED FROM THE BASIS.  THIS IS ACCOMPLISHED  1030
C       BY APPLYING THE MULTIPHASE METHOD.
C
  1000  ASSIGN 102 TO KW
        GO TO 500                                                      1035
C
C       EXECUTION OF PROGRAM BLOCK LIST
C
   102  CALL MP5(A,J1,L1,L10,JL1,Q1,KP,L3(1),ZSCHR,SSCHR)
        IF(Q1.LT.0.) GO TO 103
        FALL = 2                                                       1040
        RETURN
C
C       IF ALL COEFFICIENTS OF AN EQUALITY CONSTRAINT
C       ARE NONNEGATIVE, THEN NO FEASIBLE SOLUTION EXISTS
C                                                                      1045
   103  CALL MP2(A,J1,L2,L20,JL2,IP,ZSCHR,SSCHR,KP,Q1,N, - 1,EPS)
   120  IF(IP.NE.0) GO TO 121
        IP = L3(1)
        L30 = L30 - 1
        IF(L30.LT.1) GO TO 122                                         1050
        DO 123 S = 1,L30
   123  L3(S) = L3(S + 1)
   122  L10 = L10 - 1
        IF(L10.LT.1) GO TO 130
        DO 140 S = 1,L10                                               1055
        IF(L1(S).NE.KP) GO TO 140
        DO 141 K = S,L10
   141  L1(K) = L1(K + 1)
   140  CONTINUE
        GO TO 130                                                      1060
   121  KH = L3(1)*ZSCHR + 1
        KH1 = KH + KP*SSCHR
        Q2 = A(KH)/( - A(KH1))
        IF(Q2.GT.Q1) GO TO 124
        IP = 0                                                         1065
        GO TO 120
C
C       IP.EQ.0 MEANS THAT NO PRIOR SATISFIED CONSTRAINT CAN BE
C       VIOLATED AND THAT THE PARTICULAR CONSTRAINT CAN BE SATISFIED
```

```
C       IN ONLY ONE EXCHANGE STEP.  THIS IS ALSO POSSIBLE IF Q2          1070
C       DOES NOT EXCEED Q1.
C
  124   IF(IP.LE.M1) GO TO 130
        IF(L30.LT.1) GO TO 127
        DO 125 S = 1,L30                                                 1075
        IF(IP.NE.L3(S)) GO TO 125
        L30 = L30 − 1
        L10 = L10 − 1
        DO 126 K = S,L30
  126   L3(K) = L3(K + 1)                                                1080
  125   CONTINUE
  127   IF(L10.LT.1) GO TO 130
        DO 128 S = 1,L10
        IF(L1(S).NE.KP) GO TO 128
        DO 129 K = S,L10                                                 1085
  129   L1(K) = L1(K + 1)
  128   CONTINUE
C
  130   S = PROTO1(KP)
        PROTO1(KP) = PROTO2(IP)
        PROTO2(IP) = S                                                   1090
        CALL MP3(A,J1,0,M12,0,N,IP,KP,ZSCHR,SSCHR,1,1)
        CALL MATADR(A,N,M12,PROTO1,PROTO2,ZSCHR,SSCHR)
        IF(L30.NE.0) GO TO 1000
C                                                                        1095
C       THE FOLLOWING BLOCK DETERMINES THE CONSTRAINT FOR WHICH
C       THE ANGLE BETWEEN ITS NORMAL AND THE GRADIENT OF THE
C       OBJECTIVE FUNCTION IS MAXIMAL
C
  300   ASSIGN 301 TO KW                                                 1100
        GO TO 500
  301   CALL MP1(A,J1,0,ZSCHR,SSCHR,L1,L10,JL1)
        KP = L1(1)
        KH = L1(1)*SSCHR + 1
        IF(A(KH).GT.0..OR.Z.NE.0) GO TO 302                             1105
        KRH = R*SSCHR + 1
        IF(R.NE.0.AND.A(KRH).GT.0.) GO TO 160
        FALL = 0
        RETURN
  160   L10 = L10 + 1                                                    1110
        L1(L10) = R
        KP = R
        R = 0
        GO TO 420
C                                                                        1115
C       THE PREDEDING BLOCK REPRESENTS THE OPTIMALITY CRITERION.
C
  302   IF(J.EQ.0) GO TO 410
C
C       THE BLOCK FROM 300 UP TO THIS POINT IS ALSO USED FOR ALL         1120
C       SUBSEQUENT ITERATION STEPS.  THE NEXT BLOCK UP TO 410
C       CONTAINS THE ACTUAL DETERMINATION OF THE MAXIMAL ANGLE
C
        Q2 = 1.E + 30
        DO 303 I = 1,M12                                                 1125
        Q1 = 0.
        Q3 = 0.
        IF(L10.LT.1) GO TO 305
        DO 304 K = 1,L10
        KH = L1(K)*SSCHR + 1
        KH1 = I*ZSCHR + KH                                               1130
        Q1 = Q1 + A(KH)*A(KH1)
```

```
      304  Q3 = Q3 + A(KH1)*A(KH1)
      305  IF(Q1.GE.0..AND.Q2.LT.0.) GO TO 303
           Q1 = SIGN(Q1*Q1,Q1)/Q3                              1135
           IF(Q1.GE.Q2) GO TO 303
           Q2 = Q1
           IP = I
      303  CONTINUE
C                                                              1140
C          THE MINIMUM JUST NOW COMPUTED DETERMINES THE PIVOT ROW.
C          THE PIVOT COLUMN IS DETERMINED BY THAT MAXIMAL
C          COEFFICIENT OF THE OBJECTIVE FUNCTION WHICH CORRESPONDS
C          TO A NONZERO PIVOT ELEMENT.
C                                                              1145
           K = 1
      306  KH = IP*ZSCHR + L1(K)*SSCHR + 1
           IF(ABS(A(KH)).LT.EPS) GO TO 307
           R = L1(K)
           KP = R
           L10 = L10 - 1                                       1150
           IF(K.GT.L10) GO TO 309
           DO 308 S = K,L10
      308  L1(S) = L1(S + 1)
      309  J = 0                                               1155
           GO TO 130
      307  K = K + 1
           GO TO 306
C
C          IN THE FOLLOWING BLOCK THE VIOLATED CONSTRAINTS ARE SATISFIED   1160
C          AND (IF POSSIBLE) THE VALUE OF THE OBJECTIVE FUNCTION IS
C          INCREASED BY MEANS OF THE SIMPLEX METHOD
C
      410  IF(Z.EQ.0) GO TO 420
           CALL MP1(A,J1,Z,ZSCHR,SSCHR,L1,L10,JL1)             1165
           KP = L1(1)
           KH = Z*ZSCHR + L1(1)*SSCHR + 1
           IF(A(KH).GT.0.) GO TO 420
           KH = Z*ZSCHR + R*SSCHR + 1
           IF(A(KH).GT.0.) GO TO 411                           1170
           FALL = 2
           RETURN
      411  KP = R
           L10 = L10 + 1
           L1(L10) = R                                         1175
           R = 0
C
C          THE COLUMN DETERMINED BY THE DUOPLEX METHOD PLAYS A
C          SPECIAL ROLE.  IT IS CHOSEN AS THE PIVOT COLUMN (PROVIDED
C          THIS IS ALLOWED) ONLY IF NO OTHER PIVOT COLUMN CAN BE FOUND   1180
C
      420  CALL MP2(A,J1,L2,L20,JL2,IP,ZSCHR,SSCHR,KP,Q1,N, - 1,EPS)
           IF(IP.NE.0)GO TO 130
           IF(Z.NE.0)GO TO 421
           FALL = 1
           RETURN                                              1185
      421  IP = Z
           GO TO 130
C
C          PROCEDURE LIST                                      1190
C          LIST ENTERS ALL ROW INDICES WITH NONNEGATIVE A(I*ZSCHR)
C          INTO LIST L2 AND SETS Z EQUAL TO THE INDEX OF THE
C          FIRST ROW FOR WHICH A(I*ZSCHR) IS NEGATIVE
C
```

```
 500   L20 = 0                                                          1195
       Z = 0
       DO 501 II = 1,M12
       KH = II•ZSCHR + 1
       IF(A(KH).LT.0.) GO TO 502
       L20 = L20 + 1                                                    1200
       L2(L20) = II
       GO TO 501
 502   IF(Z.EQ.0) Z = II
 501   CONTINUE
       GO TO KW,(102,301)                                               1205
       END
C
C
C      ROUTINE FOR THE MINIMIZATION OF A LINEAR PROGRAM UNDER THE
C      ADDITIONAL CONDITION THAT ALL INDEPENDENT VARIABLES ARE          1210
C      INTEGER – VALUED.  GOMORY USES SIMPLX AND DUSEX GLOBALLY.
C      THE TABLEAU IS STORED ROW BY ROW.
C
       SUBROUTINE GOMORY (A,J1,N,M,M1,M2,M3,ZSCHR,SSCHR,FALL,
      1    PROTO1,IPROT1,PROTO2,IPROT2,L1,JL1,L2,JL2,L3,JL3,EPS)         1215
       DIMENSION A(J1),L1(JL1),L2(JL2),L3(JL3)
       INTEGER PROTO1(IPROT1),PROTO2(IPROT2),FALL,ZSCHR,SSCHR
       LOGICAL W
       N1 = N + 1
       DO 100 K = 1,N1                                                  1220
       KH = (K – 1)•SSCHR + 1
 100   A(KH) = – A(KH)
       CALL SIMPLX (A,J1,ZSCHR,SSCHR,N,M1,M2,M3,FALL,PROTO1,IPROT1,
      1    PROTO2,IPROT2,L1,JL1,L2,JL2,L3,JL3,EPS)
       IF(FALL.GT.0)RETURN                                              1225
       PRINT 101
 101   FORMAT(//• NONINTEGER OPTIMUM FOUND•)
       M4 = M1 + M2 + M3
       J = N
       NM3 = N – M3                                                     1230
       M4H = M4 + 1
       DO 102 K = 1,NM3
 110   IF(PROTO1(K).LE.N + M1 + M2.OR.PROTO1(K).GT.N + M4) GO TO 102
       PROTO1(K) = PROTO1(J)
       DO 103 I = 1,M4H                                                 1235
       KH = (I – 1)•ZSCHR + K•SSCHR + 1
       KHJ = (I – 1)•ZSCHR + J•SSCHR + 1
 103   A(KH) = A(KHJ)
       J = J – 1
       GO TO 110                                                        1240
 102   CONTINUE
       NM31 = N – M3 + 1
       DO 104 K = 1,NM31
       KH = (K – 1)•SSCHR + 1
 104   A(KH) = – A(KH)                                                  1245
 150   DO 105 I = 1,M4
       IH = I•ZSCHR + 1
       G = AINT(A(IH) + .1)
       IF((A(IH) + 0.1).LT.0.) G = G – 1.
       IF(PROTO2(I).LE.NM3.AND.ABS(G – A(IH)).GT.EPS) GO TO 202         1250
 105   CONTINUE
       RETURN
 202   IF(M4.GE.M) 106,108
 106   PRINT 107
 107   FORMAT(// • $$$$NUMERICALLY INSTABLE•)                           1255
C
C      THIS EMERGENCY EXIT IS CALLED IF THE NUMBER OF ADDITIONAL
```

```
C          CONSTRAINTS EXCEEDS THE DECLARED DIMENSION OF THE MATRIX
C
           RETURN                                                             1260
    108    M4 = M4 + 1
           PROTO2(M4) = NM3 + M4
           W = .TRUE.
           KH = M4*ZSCHR + 1
           G = AINT(A(IH))                                                     1265
           IF(A(IH).LT..0.AND.G.NE.A(IH)) G = G − 1.
           A(KH) = G − A(IH)
           DO 109 K = 1,NM3
           KHH = KH + K*SSCHR
           IHH = IH + K*SSCHR                                                  1270
           G = AINT( − A(IHH))
           IF(( − A(IHH)).LT..0.AND.G.NE.( − A(IHH))) G = G − 1.
    109    A(KHH) = − A(IHH)) − G
           CALL DUSEX (A,J1,ZSCHR,SSCHR,NM3,M4,FALL,W,PROTO1,IPROT1,
         1   PROTO2,IPROT2,L1,M4,L2,N,EPS)                                     1275
           IF(FALL.GT.0)RETURN
           GO TO 150
           END
C
C                                                                             1280
C          ROUTINE FOR DETERMINING THE MINIMUM OF A DEFINITE QUADRATIC
C          FORM SUBJECT TO LINEAR CONSTRAINTS.  BEALE USES THE SUB −
C          ROUTINES MP2,MP3,MP5,MP8, AND MP9
C
           SUBROUTINE BEALE (C,JC,A,J1,N,M,ZSCHR,SSCHR,PROTO1,IPROT1,          1285
         1   PROTO2,IPROT2,LIST,JLIST,ABLIST,JABLIS,L1,JL1,L2,JL2,FALL,EPS)
           INTEGER ZSCHR,SSCHR,FALL,PROTO1,PROTO2,ABLIST,V,S,Z,T,R
           DIMENSION C(JC),A(J1),PROTO1(IPROT1),PROTO2(IPROT2),LIST(JLIST),
         1   ABLIST(JABLIS),L1(JL1),L2(JL2)
           REAL MAX                                                           1290
           LOGICAL B1
           DO 1 K = 1,N
           L1(K) = K
      1    PROTO1(K) = K
           IF(M.EQ.0) GO TO 4                                                 1295
           DO 2 I = 1,M
           L2(I) = I
      2    PROTO2(I) = N + I
      4    L10 = N
           L20 = M                                                            1300
           DO 3 K = 1,N
      3    ABLIST(K) = 0
           LIST0 = 0
           M1 = M
C                                                                             1305
C          THE BLOCK 1XXX DETERMINES THE PIVOT COLUMN KP
C
   1000    IF(LIST0.EQ.0) GO TO 1001
           CALL MP8 (C,JC,0,0,1,LIST,LIST0,JLIST,KP,MAX)
           IF(ABS(MAX).GT.1.E − 8) GO TO 2000                                 1310
   1001    IF(L10.LE.0) GO TO 1002
           CALL MP5(C,JC,L1,L10,JL1,Q1,KP,0,0,1)
           IF(Q1.LT. − EPS) GO TO 2000
   1002    FALL = 0
           RETURN                                                             1315
C
C          IF ALL LEADING ELEMENTS OF THE U − COLUMNS EQUAL ZERO
C          AND NO LEADING ELEMENT OF AN X − COLUMN IS NEGATIVE,
C          THE MINIMUM HAS BEEN OBTAINED
C                                                                             1320
```

```
C        THE BLOCK 2XXX DETERMINES THE PIVOT ROW
C
 2000    KH = KP*(N + 1) − ((KP − 1)*KP)/2 + 1
         MAX = 0.
         IF(C(KH).GT.EPS) MAX = C(KP + 1)/C(KH)                          1325
         V = − 1
         IF(MAX.GT.0.)V = 1
         CALL MP2(A,J1,L2,L20,JL2,IP,ZSCHR,SSCHR,KP,Q1,N,V,EPS)
         IF(IP.NE.0.OR.MAX.NE.0.) GO TO 3000
         FALL = 1                                                        1330
         RETURN
C
C        IF IP AND MAX EQUAL ZERO, THE SOLUTION IS UNBOUNDED
C
 3000    IF(IP.EQ.0.OR.Q1.GT.ABS(MAX)) GO TO 4000                        1335
         I = PROTO1(KP)
         PROTO1(KP) = PROTO2(IP)
         PROTO2(IP) = I
         CALL MP3(A,J1,1,M,0,N,IP,KP,ZSCHR,SSCHR,1,1)
         B1 = .TRUE.                                                     1340
         CALL MP9(A,J1,C,JC,IP,KP,N,M1,ZSCHR,SSCHR,B1)
         GO TO 6000
C
C        IF Q1 DOES NOT EXCEED MAX, THE PIVOT ROW IS
C        CONTAINED IN THE CONSTRAINT SECTION                             1345
C
 4000    ABLIST(KP) = ABLIST(KP) + 1
         T = 0
         R = 0
 4001    IF(LIST0.EQ.0) GO TO 5000                                       1350
         DO 4002 I = 1,LIST0
         IF(LIST(I).EQ.KP) GO TO 4003
 4002    CONTINUE
         GO TO 5000
 4003    Z = 0                                                          1355
         KKP = KP + 1
         DO 4004 S = 1,KKP
         IS = S − 1
         KH = T*ZSCHR + IS*SSCHR + 1
         KH1 = KP + Z + 1                                                1360
         A(KH) = C(KH1)
 4004    Z = Z + N − IS
         NKP = N − KP
         IF(NKP.LT.1) GO TO 4005
         DO 4006 S = 1,NKP                                               1365
         KH = T*ZSCHR + (KP + S)*SSCHR + 1
         KH1 = KP*(N + 1) − (KP*(KP − 1))/2 + S + 1
 4006    A(KH) = C(KH1)
 4005    IP = T
         CALL MP3(A,J1,R,M,0,N,IP,KP,ZSCHR,SSCHR,1,1)                    1370
         B1 = .FALSE.
         CALL MP9(A,J1,C,JC,IP,KP,N,M1,ZSCHR,SSCHR,B1)
         CALL MATCDR(C,N,PROTO1)
         CALL MATADR(A,N,M,PROTO1,PROTO2,ZSCHR,SSCHR)
         GO TO 1000                                                      1375
C
 5000    LIST0 = LIST0 + 1
         LIST(LIST0) = KP
         PROTO2(M + 1) = PROTO1(KP)
         PROTO1(KP) = N + M1 + KP                                        1380
         L2(M + 1) = M + 1
         L20 = M + 1
         IF(L10.LT.1) GO TO 5004
```

```
      DO 5001 S = 1,L10
5001  IF(L1(S).EQ.KP) T = S                                        1385
      L10 = L10 − 1
5004  IF(L10.LT.T.OR.T.LE.0) GO TO 5006
      DO 5003 S = T,L10
5003  L1(S) = L1(S + 1)
5006  M = M + 1                                                    1390
      T = M
      R = 1
      GO TO 4001
6000  IF(PROTO2(IP).LE.N + M1) GO TO 6050
      PROTO2(IP) = PROTO2(M)                                       1395
      L20 = L20 − 1
      L10 = L10 + 1
      L1(L10) = KP
      LIST0 = LIST0 − 1
      IF(LIST0.LT.1) GO TO 6004                                    1400
      DO 6001 K = 1,LIST0
      IF(LIST(K).EQ.KP) GO TO 6003
6001  CONTINUE
      GO TO 6004
6003  DO 6005 I = K,LIST0                                          1405
6005  LIST(I) = LIST(I + 1)
6004  NN = N + 1
      KH1 = IP*ZSCHR + 1
      KH2 = M*ZSCHR + 1
      DO 6002 K = 1,NN                                             1410
      KH0 = (K − 1)*SSCHR
      KH3 = KH1 + KH0
      KH4 = KH2 + KH0
6002  A(KH3) = A(KH4)
      M = M − 1                                                    1415
      ABLIST(KP) = ABLIST(KP) − 1
6050  CALL MATCDR(C,N,PROTO1)
      CALL MATADR(A,N,M,PROTO1,PROTO2,ZSCHR,SSCHR)
      GO TO 1000
      END                                                          1420
C
C
C     ROUTINE FOR COMPUTING THE MINIMUM OF A QUADRATIC
C     FORM SUBJECT TO LINEAR CONSTRAINTS BY THE LONG FORM
C     OF THE WOLFE METHOD.  WOLFE USES THE SUBROUTINES             1425
C     MP2,MP3, AND MP5
C
      SUBROUTINE WOLFE (A,J1,X,X0,JX,ZSCHR,SSCHR,N,M,PROTO1,
     1    IPROT1,PROTO2,IPROT2,FALL,L1,JL1,L2,JL2,LIST1,JLIST1,EPS)
      DIMENSION A(J1),L1(JL1),L2(JL2),X(JX),LIST1(JLIST1)          1430
      INTEGER FALL,PROTO1(IPROT1),PROTO2(IPROT2),S,T,ZSCHR,SSCHR,V
C
C     INITIALIZATION OF THE INDICES AND INDEX LISTS
C
      N1 = 3*N + M + 1                                             1435
      DO 1 K = 1,N1
1     PROTO1(K) = K
      NM = N + M
      NM3 = NM + 3
      DO 2 I = 1,NM                                                1440
      PROTO2(I) = N1 + I
2     L2(I) = I
      L20 = NM
      DO 3 K = 1,N
3     L1(K) = K                                                    1445
      NN = 2*N + M + 1
```

```
        NN1 = N1 − 1
        DO 4 K = NN,NN1
        J = K − NM
   4    L1(J) = K                                                    1450
        L10 = 2*N
        V = − 1
C
C       PHASE 1 OF THE WOLFE METHOD
C       PROCEDURE FOR COMPUTING THE AUXILIARY OBJECTIVE FUNCTION      1455
C
        PRINT 10
  10    FORMAT(//* WOLFE 1*)
1000    NNN = N + 1
        DO 1001 KK = 1,NNN                                           1460
        K = KK − 1
        Q1 = 0.
        KH = K*SSCHR + 1
        DO 1002 I = 1,M
        KH0 = KH + I*ZSCHR                                           1465
1002    Q1 = Q1 + A(KH0)
1001    A(KH) = Q1
        DO 1003 K = NNN,N1
        KH = K*SSCHR + 1
1003    A(KH) = 0.                                                   1470
C
C       COMPUTATION OF THE MINIMUM OF THE AUXILIARY OBJECTIVE
C       FUNCTION BY THE SIMPLEX METHOD
C
1100    CALL MP5(A,J1,L1,L10,JL1,Q1,KP,0,ZSCHR,SSCHR)               1475
        IF(Q1.GE.0.) GO TO 1200
C
C       IF Q1 IS NONNEGATIVE, THE MINIMUM OF THE AUXILIARY
C       OBJECTIVE FUNCTION HAS BEEN OBTAINED.
C                                                                    1480
        CALL MP2(A,J1,L2,L20,JL2,IP,ZSCHR,SSCHR,KP,Q1,N1,V,EPS)
        IF(IP.NE.0) GO TO 1101
        FALL = 2
        RETURN
1101    I = PROTO1(KP)                                               1485
        PROTO1(KP) = PROTO2(IP)
        PROTO2(IP) = I
        CALL MP3(A,J1,0,NM,0,N1,IP,KP,ZSCHR,SSCHR,1,1)
        CALL MATADR(A,ZSCHR − 1,NM,PROTO1,PROTO2,ZSCHR,SSCHR)
        GO TO 1100                                                   1490
C
C       PROCEDURE FOR DELETION OF THE W − AND Z − COLUMNS NOT
C       IN THE BASIS
C
1200    LIST10 = 0                                                   1495
        I = 0
        DO 1201 K = 1,NN1
1204    IF(K.GT.NN1 − I) GO TO 1203
        IF(.NOT.(PROTO1(K).GE.NN.AND.PROTO1(K).LE.3*N + M.OR.PROTO1(K).GE.
   1    3*N + M + 2.AND.PROTO1(K).LE.4*N + 2*M + 1)) GO TO 1201      1500
        LIST10 = LIST10 + 1
        LIST1(LIST10) = K + I
        NNN = N1 − 1 − I
        DO 1202 S = K,NNN
1202    PROTO1(S) = PROTO1(S + 1)                                    1505
        I = I + 1
        GO TO 1204
1201    CONTINUE
```

```
 1203   ASSIGN 2000 TO KW
        GO TO 9000                                                          1510
C
C       EXECUTION OF MP11
C
C       START OF THE SECOND PHASE OF WOLFE'S METHOD.  COMPUTATION
C       OF THE SECOND AUXILIARY OBJECTIVE FUNCTION                          1515
C
 2000   PRINT 11
   11   FORMAT(//* WOLFE 2*)
        CALL MATADR(A,N1,NM,PROTO1,PROTO2,ZSCHR,SSCHR)
        NNN = N1 + 1                                                        1520
        DO 2001 KK = 1,NNN
        K = KK − 1
        Q1 = 0.
        KH0 = K*SSCHR + 1
        DO 2002 I = 1,NM                                                    1525
        KH = KH0 + I*ZSCHR
 2002   IF(PROTO2(I).GE.2*N + M + 1.AND.PROTO2(I).LE.3*N + M.OR.PROTO2(I).
      1    GE.3*N + 2*M + 2.AND.PROTO2(I).LE.4*N + 2*M + 1)
      2    Q1 = Q1 + A(KH)
 2001   A(KH0) = Q1                                                         1530
C
C       DETERMINATION OF THE MINIMUM OF THE SECOND AUX. OBJ. FUNCTION
C
 2100   NN1 = N1 − 1
        ASSIGN 2200 TO KW                                                   1535
        GO TO 8000
C
C       EXECUTION OF MP12
C
 2200   CALL MATADR(A,N1,NM,PROTO1,PROTO2,ZSCHR,SSCHR)                      1540
        CALL MP5(A,J1,L1,L10,JL1,Q1,KP,0,ZSCHR,SSCHR)
        IF(Q1.GE.0.) GO TO 2300
        CALL MP2(A,J1,L2,L20,JL2,IP,ZSCHR,SSCHR,KP,Q1,N1,V,EPS)
        IF(IP.NE.0) GO TO 2201
        FALL = 2                                                            1545
        RETURN
 2201   I = PROTO1(KP)
        PROTO1(KP) = PROTO2(IP)
        PROTO2(IP) = I
        CALL MP3(A,J1,0,NM,0,N1,IP,KP,ZSCHR,SSCHR,1,1)                      1550
        GO TO 2100
C
C       DELETION OF THE Z − COLUMNS FROM THE TABLEAU
C
 2300   I = 0                                                               1555
        LIST10 = 0
        NN1 = N1 − 1
        DO 2301 K = 1,NN1
 2306   IF(K.GT.NN1 − I) GO TO 2305
        IF(.NOT.(PROTO1(K).GE.2*N + M + 1.AND.PROTO1(K).LE.3*N + M.OR.      1560
      1    PROTO1(K).GE.3*N + 2*M + 2.AND.PROTO1(K).LE.4*N + 2*M + 1)) GO TO 2301
        LIST10 = LIST10 + 1
        LIST1(LIST10) = K + I
        NNN = N1 − I − 1
        DO 2302 S = K,NNN                                                   1565
 2302   PROTO1(S) = PROTO1(S + 1)
        I = I + 1
        GO TO 2306
 2301   CONTINUE
 2305   ASSIGN 2303 TO KW                                                   1570
        GO TO 9000
```

```
C
C         EXECUTION OF MP11
C
 2303     PRINT 13                                                            1575
   13     FORMAT(//* WOLFE 3*)
          CALL MATADR(A,N1,NM,PROTO1,PROTO2,ZSCHR,SSCHR)
          DO 2304 KK = 1,N1
          K = KK − 1
          KH = K*SSCHR + 1                                                    1580
 2304     A(KH) = 0.
          KH = N1*SSCHR + 1
          A(KH) = − 1.
C
C         START OF THE THIRD PHASE OF WOLFE'S METHOD                          1585
C
 3000     NN1 = N1
          ASSIGN 3200 TO KW
          GO TO 8000
C                                                                            1590
C         EXECUTION OF MP12
C
 3200     X0 = A(1)
          DO 3204 I = 1,NM
 3204     X(I) = 0.                                                           1595
          DO 3201 I = 1,NM
          KH = I*ZSCHR + 1
          KH0 = PROTO2(I)
 3201     IF(KH0.LE.N)X(KH0) = A(KH)
          CALL MP5(A,J1,L1,L10,JL1,Q1,KP,0,ZSCHR,SSCHR)                      1600
          IF(Q1.LT.0.) GO TO 3202
 3203     FALL = 2
          RETURN
 3202     CALL MP2(A,J1,L2,L20,JL2,IP,ZSCHR,SSCHR,KP,Q1,N1,V,EPS)
          IF(IP.EQ.0) GO TO 3203                                             1605
          I = PROTO1(KP)
          PROTO1(KP) = PROTO2(IP)
          PROTO2(IP) = I
          CALL MP3(A,J1,0,NM,0,N1,IP,KP,ZSCHR,SSCHR,1,1)
          CALL MATADR(A,N1,NM,PROTO1,PROTO2,ZSCHR,SSCHR)                     1610
          IF(− A(1).LT.1.) GO TO 3000
C
C         COMPUTATION OF THE SOLUTION
C
 3300     Q2 = A(1) − X0                                                      1615
          Q1 = (A(1) + 1.)/Q2
          Q2 = − (1. + X0)/Q2
          DO 3301 I = 1,N
 3301     X(I) = Q1*X(I)
          DO 3302 I = 1,NM                                                    1620
          KH = PROTO2(I)
          KH0 = I*ZSCHR + 1
 3302     IF(KH.LE.N)X(KH) = X(KH) + Q2*A(KH0)
          FALL = 0
          RETURN                                                             1625
C
C         PROCEDURE MP11
C         ROUTINE FOR REMOVAL OF THE COLUMNS OF
C         TABLEAU A WHICH ARE CONTAINED IN LIST1
C                                                                            1630
 9000     IT = 0
          MM = NM + 1
          KE = N1 − 1
          IF(LIST10.LT.1) GO TO 9001
```

```
        DO 9002 K = 1,LIST10                                        1635
        KA = LIST1(K)
        DO 9003 S = KA,KE
        DO 9003 I = 1,MM
        KH = (I − 1)•ZSCHR + (S − IT)•SSCHR + 1
        KH0 = KH + SSCHR                                            1640
9003    A(KH) = A(KH0)
9002    IT = IT + 1
9001    N1 = N1 − LIST10
        GO TO KW,(2000,2303)
C                                                                   1645
C       PROCEDURE MP12
C       GENERATION OF LIST L1 OF ALL COLUMNS ADMISSIBLE AS PIVOT COLUMNS
C
8000    L10 = 0
        DO 8001 K = 1,NN1                                           1650
        IF(PROTO1(K).GT.N) GO TO 8002
        DO 8003 I = 1,NM
        IF(PROTO1(K) + N.EQ.PROTO2(I)) GO TO 8001
8003    CONTINUE
8005    L10 = L10 + 1                                              1655
        L1(L10) = K
        GO TO 8001
8002    IF(PROTO1(K).GT.2•N) GO TO 8005
        DO 8004 I = 1,NM
        IF(PROTO1(K) − N.EQ.PROTO2(I)) GO TO 8001                  1660
8004    CONTINUE
        GO TO 8005
8001    CONTINUE
        GO TO KW,(2200,3200)
        END                                                        1665
C
```

INDEX